Lecture Notes in Mathematics

Edited by A. Dold and B. Eckmann

984

Antonio Bove
Jeff E. Lewis
Cesare Parenti

Propagation of Singularities for Fuchsian Operators

Springer-Verlag
Berlin Heidelberg New York Tokyo 1983

Authors

Antonio Bove
Department of Mathematics, University of Trento
38050 Povo (Trento), Italy

Jeff E. Lewis
Department of Mathematics, University of Illinois at Chicago
P.O.Box 4348, Chicago, IL 60680, USA

Cesare Parenti
Department of Mathematics, University of Bologna
Piazza di Porta S. Donato, 5, 40127 Bologna, Italy

AMS Subject Classifications (1980): 58G16, 58G17, 35L40

ISBN 3-540-12285-0 Springer-Verlag Berlin Heidelberg New York Tokyo
ISBN 0-387-12285-0 Springer-Verlag New York Heidelberg Berlin Tokyo

Printing and binding: Beltz Offsetdruck, Hemsbach/Bergstr.
2146/3140-543210

TABLE OF CONTENTS

	Introduction	1
1.	Preliminaries and Review of Results of N. Hanges	10
2.	General Fuchsian Systems	24
3.	Applications to Fuchsian Hyperbolic P.D.E.	96
4.	Operators with Multiple Non-Involutive Characteristics	133
5.	References	158
6.	Subject Index	161

ACKNOWLEDGEMENTS

One of us (J.E.L.) would like to thank the Italian Research Council for supporting his staying at the University of Bologna during the completion of this work. A.B. and C.P. were partially supported by the C.N.R., gruppo G.N.A.F.A.

The authors would like to thank Mrs S. Serra and Mrs. M. Stettermajer for their excellent typing work.

INTRODUCTION

The main purpose of this monograph is the study of Fuchsian systems of the form

$$Pu = (t\partial_t I_N - A(t,x,D_t,D_x))u(t,x) = f(t,x) \ , \tag{0.1}$$

where A is an $N \times N$ matrix of classical pseudodifferential operators (pdo) of order 0 defined on $\mathbb{R}^{n+1} = \mathbb{R}_t \times \mathbb{R}^n_x$. More precisely, we are interested in describing C^∞-singularities of the solutions of system (0.1) i.e. the set $WF(u) \setminus WF(Pu)$, where $WF(v)$ denotes the wave front set of the distribution v as defined in L. Hörmander [14] (for a vector-valued distribution $v = (v_1,\dots,v_N) \in \mathcal{D}'^N$ we put $WF(v) = \bigcup_{j=1}^{N} WF(v_j)$).

It is well known that the structure of the set $WF(u) \setminus WF(Pu)$ depends on the characteristics of the operator P, i.e.

$$WF(u) \setminus WF(Pu) \subset \{(t,x,\tau,\xi) \in T^* \mathbb{R}^{n+1} \setminus 0 \ \mid \ t\tau = 0\} = \mathrm{Char}\ P \tag{0.2}$$

Near a point $\rho_0 = (t_0,x_0,\tau^0,\xi^0) \in \mathrm{Char}\ P$ for which $t_0 \neq 0$ or $t_0 = 0$ and $\tau^0 \neq 0$, $\xi^0 \neq 0$, a complete description of $WF(u) \setminus WF(Pu)$ follows from the general results of J.J. Duistermaat - L. Hörmander [10]; in particular $WF(u) \setminus WF(Pu)$ is invariant under the action of the hamiltonian vector fields $H_\tau = \frac{\partial}{\partial t}$ and $H_t = -\frac{\partial}{\partial \tau}$ respectively.

Therefore we concentrate our analysis of $WF(u) \setminus WF(Pu)$ near the points of the two following disjoint subsets of Char P :

$$(0.3)\qquad \begin{aligned} \Sigma &= \{(t,x,\tau,\xi) \in T^* \mathbb{R}^{n+1} \setminus 0 \mid t = 0,\ \tau = 0\} \\ \dot{N}^* \mathbb{R}^n &= \{(t,x,\tau,\xi) \in T^* \mathbb{R}^{n+1} \mid t = 0,\ \xi = 0,\ \tau \neq 0\} \end{aligned}$$

In order to study singularities near a point $\rho_o \in \Sigma \cup \dot{N}^* \mathbb{R}^n$ a general policy consists in constructing a left microlocal parametrix for the system (0.1) i.e. an operator $E : \mathcal{D}'(\mathbb{R}^{n+1})^N \to \mathcal{D}'(\mathbb{R}^{n+1})^N$ such that $EP - I_N$ is a smoothing operator in a conical neighborhood of ρ_o. However it does not seem to be an easy task to construct parametrices directly for system (0.1). It turns out that it is more convenient to reduce (0.1) to a simpler equivalent canonical form and then construct a parametrix for the simplified system.

In carrying over this program an important role is played by the quantity:

$$(0.4)\qquad I^P_{\rho_o}(\zeta) = \det(\zeta I_N - A_o(\rho_o)),\ \zeta \in \mathbb{C},$$

where A_o denotes the principal symbol of the matrix A . The polynomial (0.4) shall be called the *indicial polynomial* of the system (0.1) at the point $\rho_o \in \Sigma \cup \dot{N}^* \mathbb{R}^n$. We point out that if (0.1) is an ordinary differential system (i.e. if A is a matrix of functions) then the polynomial $I^P_{\rho_o}(\zeta)$ coincides with the classical indicial polynomial encountered in the theory of ordinary differential systems with regular singularities (see e.g. B. Malgrange [21]). The precise canonical form of the system (0.1) in a conic neighborhood of ρ_o depends on whether the roots $\zeta_1(\rho_o),\dots,\zeta_N(\rho_o)$ of the *indicial equation* $I^P_{\rho_o}(\zeta) = 0$ differ by non zero integers. We shall say that the *Fuchs condition* $(F)_{\rho_o}$ is satisfied at ρ_o if

$$(F)_{\rho_o} : \zeta_i(\rho_o) - \zeta_j(\rho_o) \notin \mathbb{Z} \setminus \{0\} , \quad i,j = 1,\ldots,N.$$

We now describe the results we obtain in the two different cases $\rho_o \in \Sigma$ and $\rho_o \in \dot{N}^* \mathbb{R}^n$.

a) Case $\rho_o \in \Sigma$

Let us first suppose that the Fuchs condition $(F)_{\rho_o}$ is satisfied by the system (0.1). It will be proved in Section 2 that there exists an intertwining matrix Q of classical p d o's of order 0, elliptic near ρ_o, such that in a conic neighborhood of ρ_o:

$$(0.5) \qquad Q^{-1} P Q \equiv \tilde{P} = t\partial_t I_N - \tilde{A}(x,D_x) ,$$

where $\tilde{A}$ is a $N \times N$ matrix of p d o's of order 0 which does not depend on t and D_t.

Note that $I^P_{\rho_o}(\zeta) = I^{\tilde{P}}_{\rho_o}(\zeta)$. We are thus reduced to analyze the singularities near ρ_o of the solutions to the system

$$(0.6) \qquad \tilde{P}u = (t\partial_t I_N - \tilde{A}(x,D_x))v(t,x) = g(t,x)$$

with $v = Q^{-1}u$, $g = Q^{-1}f$.

In the case $N = 1$ explicit microlocal parametrices for equation (0.6) have been constructed by N. Hanges [13]; these constructions may be readily extended to give parametrices for (0.6) in the general case $N \geq 1$. The results are reviewed in Section 1. As a consequence the following theorem can be proved.

THEOREM 1. Let $\rho_o = (0,x_o,0,\xi^o)$, $\xi^o \neq 0$, and define

$$(0.7) \qquad \begin{cases} \gamma_1^{\pm} = \{(0,x_o,s,\xi^o) \in T^* \mathbb{R}^{n+1} \setminus 0 \mid \pm s > 0\} \\ \gamma_2^{\pm} = \{(s,x_o,0,\xi^o) \in T^* \mathbb{R}^{n+1} \setminus 0 \mid \pm s > 0\} \end{cases}$$

Let $Pu = f \in \mathcal{D}'(\mathbb{R}^{n+1})^N$ and suppose that $\rho_o \notin WF(f)$. Then

i) If for every $j \in \{1,2\}$ there is a choice of the sign + or - for which $\gamma_j^{\pm}$ does not intersect $WF(u)$ then $\rho_o \notin WF(u)$.

ii) If $I^P_{\rho_o}(\zeta) \neq 0$ for $\zeta = 0,1,2,\ldots$ and both γ_1^+ and γ_1^- do not intersect $WF(u)$ then $\rho_o \notin WF(u)$.

iii) If $I^P_{\rho_o}(\zeta) \neq 0$ for $\zeta = -1,-2,-3,\ldots$ and both γ_1^+ and γ_1^- do not intersect $WF(u)$ then $\rho_o \notin WF(u)$.

For $N=1$ this theorem was proved by N. Hanges [13]; for a less constructive proof see V.Ya. Ivrii [16], R. Melrose [22], and for a slightly different case see S. Alinhac [3].

We remark that in Section 1 we give some results on the structure of the solution u when the indicial polynomial has integer roots (for this case see also [16]). If the Fuchs condition $(F)_{\rho_o}$ is not satisfied the situation is more involved since we find a non-trivial obstruction to the reduction (0.5). In the absence of the Fuchs condition me make a tricky reduction inspired by M. Kashiwara - T. Oshima [19] and T. Ôaku [26] and an enlargement of the original system (0.1) to show that, if $Pu = f$, then there are vectors U,F with $WF(u) = WF(U)$, $WF(F) = WF(f)$ near ρ_o which satisfy near ρ_o a suitable Fuchsian system:

$$P'U = (t\partial_t I_M - A'(x,D_x))U = F \tag{0.8}$$

where M is larger than N, A' is a $M\times M$ matrix of pdo's of order 0 which do not depend on t and D_t and $I^{P'}_{\rho_o}(\zeta)$ has the same roots as $I^P_{\rho_o}(\zeta)$. The precise construction of the system P' is quite technical and forms the core of Section 2 to which we refer the reader for further details.

As a consequence Theorem 1 holds for any system of the form (0.1).

For results of the same kind in the hyperfunction or analytic setting see M. Kashiwara,

T, Kawai, T. Oshima [18], H. Tahara [28] and the recent work of T. Ôaku [26].

b) Case $\rho_o \in \dot{N}^* \mathbb{R}^n$

The last part of Section 2 is dedicated to finding a canonical form for the system (0.1) near points $\rho_\pm = (0, x_o, \tau = \pm 1, 0)$ of the conormal bundle of $\mathbb{R}^n_x$. If the Fuchs condition $(F)_{\rho_\pm}$ is satisfied at $\rho_\pm$, we show, as in N. Hanges [12], that the system P is microlocally equivalent near $\rho_\pm$ to the multiplication operator tI_N. Otherwise, using an idea of M. Kashiwara - T. Oshima [19], we show that (0.1) may be reduced via an elliptic intertwining operator to the canonical form

(0.9) $$P_\pm = t\partial_t I_N - B_\pm(x, D_t, D_x) .$$

Furthermore, $B_\pm$ is block upper triangular and $b_{ii,\pm}(x, D_t, D_x) = b_{ii,\pm}(x)$. As a consequence we obtain the following result.

THEOREM 2. Consider the operator P defined in (0.1) as a mapping on $\mathcal{M}^N_{\rho_\pm}$, where $\mathcal{M}_{\rho_\pm}$ is the stalk over $\rho_\pm$ of the sheaf of microdistributions in $\mathbb{R}^{n+1}$. Then:

i) P is surjective, i.e. given $f \in \mathcal{D}'(\mathbb{R}^{n+1})^N$ there is a $u_\pm \in \mathcal{D}'(\mathbb{R}^{n+1})^N$ such that $\rho_\pm \notin WF(Pu_\pm - f)$

ii) Ker P is isomorphic to N copies $\mathcal{D}'_{x_o}(\mathbb{R}^n)$, the space of germs at x_o of distributions in the x variable.

When the Fuchs condition is satisfied and the roots of the indicial equation $I^P_{\rho_\pm}(\zeta) = 0$ are simple, more precise information on the structure of $\ker P \subset \mathcal{M}^N_{\rho_\pm}$ can be obtained (see the remark following Theorem 2.3 of Section 2). For closely related results in the hyperfunction setting see M. Kashiwara - T. Oshima [19] and H. Tahara [28].

In Section 3 we apply the previously developed theory of Fuchsian systems to the study of a scalar Fuchsian operator of order m and weight $m-k$, $1 \leq k \leq m$:

(0.11) $$P(t,x,D_t,D_x) = t^k P_m + t^{k-1} P_{m-1} + \dots + P_{m-k},$$

where $P_{m-j}(t,x,D_t,D_x)$ is a classical (pseudo) differential operator of order $m-j$ on $\mathbb{R}^{n+1}$, $j = 0,\dots,k$.

Operators of the form (0.11) are a particular case of those considered by M. Baouendi-C. Goulaouic [6]. The literature on Fuchsian differential operators is extensive. Besides the works cited previously and in the text, we mention some recent work on Fuchsian operators. Parabolic Fuchsian operators have been studied by C. Baiocchi-M. Baouendi [5] and elliptic Fuchsian operators are studied by P. Bolley - J. Camus [7]. Second order elliptic operators of Fuchs type are extensively treated in F. Treves [30]. For results on existence when $P_m(t,x,D_t,D_x)$ is strictly hyperbolic with respect to t see S. Alinhac [2] and N. Hanges [12]. J. F. Nourrigat [25] studies boundary value problems for strictly hyperbolic Fuchsian operators. G. Roberts [27] proves a Calderon type uniqueness theorem for operators of the form (0.11).

Here we assume that the hyperplane $t = 0$ is not characteristic for P_m and that the principal symbol p_m of P_m admits the factorization:

(0.12) $$p_m(t,x,\tau,\xi) = q(t,x,\tau,\xi)^r \, e(t,x,\tau,\xi) ,$$

where e is an elliptic factor, $q(t,x,\tau,\xi) = \tau - \lambda(t,x,\xi)$ with λ a smooth real function positively homogeneous of degree 1 in ξ and r is a positive integer. If $r \geqq 2$ we assume that the operator P satisfies a Levi condition with respect to the factor q (see J. Chazarain [8]).

Firstly we are interested in studying the singularities of a solution u of the equation $Pu = f$ near points $\rho_0 \in T^* \mathbb{R}^{n+1} \setminus 0$ of the form $\rho_0 = (0,x_0,\tau = \lambda(0,x_0,\xi^0),\xi^0)$, $\xi^0 \neq 0$.

Note that the particular form of (0.11) implies that P satisfies a Levi condition with respect to the factor t if $k \geq 2$. Since the Poisson bracket $\{t,q\} = -1$, points ρ_o belong to the non-involutory manifold

$$\hat{\Sigma} = \{(t,x,\tau,\xi) \in T^* \mathbb{R}^{n+1} \setminus 0 \mid t = 0,\ \tau = \lambda(0,x,\xi), \xi \neq 0\} \tag{0.13}$$

which is characteristic for P.

In Section 3 we show that the Fuchs equation $Pu = f$ can be microlocally transformed into a Fuchsian system of the form (0.1) (with $N = \max(k,r)$) whose indicial polynomial at ρ_o can be explicitly computed in terms of the principal symbols $p_{m-j}(\rho_o)$ of the operators P_{m-j}, $j = 0,\ldots,k$.

An application of Theorem 1 gives the following

THEOREM 3. Let P be as in (0.11) and satisfy the above hypotheses. Let $\rho_o \in \hat{\Sigma} \setminus WF(Pu)$, $u \in \mathcal{D}'(\mathbb{R}^{n+1})$. Denote by $\gamma_1(s)$ (resp. $\gamma_2(s)$) the integral curve of the hamiltonian field H_t (resp. H_q) with $\gamma_1(0) = \rho_o$ (resp. $\gamma_2(0) = \rho_o$) and let $\gamma_j^{\pm}$, $j = 1,2$ be the four open half-bicharacteristic curves with $\pm s > 0$. Then:

i) If for each $j \in \{1,2\}$ there is a choice of the sign + or − for which $\gamma_j^{\pm} \cap WF(u) = \emptyset$ then $\rho_o \notin WF(u)$.

ii) Suppose that $r \leq k$ and the roots of the indicial polynomial of the associated system do not lie in $\{0,1,2,\ldots\}$; if γ_1^+ and γ_1^- do not intersect $WF(u)$, then $\rho_o \notin WF(u)$.

iii) Suppose that $r \geq k$ and that the roots of the indicial polynomial of the associated system do not lie in $\{-1,-2,-3,\ldots\}$; if γ_2^+ and γ_2^- do not intersect $WF(u)$, then $\rho_o \notin WF(u)$.

The conclusions of part ii) and iii) of Theorem 3 are sharp, for $k \neq r$, in a sense clarified by the remarks following the proof of Theorems 3.1 and 3.2 of Section 3.

We conclude Section 3 by studying a Fuchsian differential operator of the form (0.11) near points $\rho_{\pm} = (0,x_o,\tau=\pm 1,\xi=0)$. Again we reduce P to an equivalent Fuchsian system of dimension k. The indicial polynomial of the so obtained system coincides with the classical indicial polynomial as defined in M. Baouendi - C. Goulaouic [6]. We then apply the results of Section 2 for Fuchsian systems to describe microlocally the kernel and the cokernel of P.

In Section 4 we come to the original motivation for this work: the study of propagation of singularities for (pseudo) differential operators with multiple non-involutive characteristics. Let

(0.14) $$P = P_m + P_{m-1} + \dots$$

and for $j=1,2$, we let $q_j(t,x,\tau,\xi) = \tau - \lambda_j(t,x,\xi)$, where $\lambda_j(t,x,\xi)$ is a smooth real function positively homogeneous of degree 1 in ξ. Denote by $\Sigma_j = \{(t,x,\tau,\xi) \mid q_j(t,x,\tau,\xi) = 0\}$ and let $\Sigma = \Sigma_1 \cap \Sigma_2$. We suppose that near Σ

(0.15) $$p_m(t,x,\tau,\xi) = (q_1(t,x,\tau,\xi))^k \, (q_2(t,x,\tau,\xi))^r \, e(t,x,\tau,\xi) ,$$

where $e(t,x,\tau,\xi)$ is an elliptic symbol homogeneous of degree $m-(k+r)$ in (τ,ξ). Our second assumption is

(0.16) $$\begin{cases} \text{if } k \geqq 2, P \text{ satisfies the Levi condition with respect to } q_1. \\ \text{if } r \geqq 2, P \text{ satisfies the Levi condition with respect to } q_2. \end{cases}$$

The third assumption is that Σ is symplectic, i.e.:

(0.17) $$\{q_1,q_2\} \neq 0 \quad \text{on } \Sigma .$$

Under the above assumptions we show that if, e.g., $\{q_1,q_2\}(\rho_o) < 0$, then P is microlocally equivalent near Σ to a Fuchsian hyperbolic operator of order m and weight $m-k$. We can now state the main propagation result. Let $\rho_o \in \Sigma$ and for $j=1,2$, let $\gamma_j(s)$ be the integral curve of the hamiltonian field H_{q_j} such that $\gamma_j(0) = \rho_o$. Denote by $\gamma_j^{\pm} = \{\gamma_j(s) \mid \pm s > 0\}$ the four open half-bicharacteristics through ρ_o.

THEOREM 4. Let P be as in (0.14) and satisfy (0.15), (0.16), (0.17). Let $\rho_o \in \Sigma$ and $\rho_o \notin WF(Pu)$. If for every $j \in \{1,2\}$ there is a choice of the sign + or − such that $\gamma_j^{\pm} \cap WF(u) = \emptyset$, then $\rho_o \notin WF(u)$.

To obtain finer results on propagation of singularities for (0.14) corresponding to parts ii) and iii) of Theorem 3, one needs to know the coefficients of the indicial polynomial. Following the coefficients of the indicial polynomial does not seem to be an easy task due to the complexity of the Levi conditions; we give a complete detailed description in the case $1 \leq k,\ r \leq 2$, which includes the previous results of N. Hanges [13].

1. PRELIMINARIES AND REVIEW OF RESULTS OF N. HANGES

We state here some extensions to the vector-valued situation of results of N. Hanges [13].

To fix our notation let X be an open subset of $\mathbb{R}^n$ and let $\tilde{X} = (-T,T) \times X$ be a neighborhood of $\{0\} \times X$ in $\mathbb{R}^{n+1}$. The points of $T^*\tilde{X}$ will be denoted by (t,x,τ,ξ), $(t,x) \in \tilde{X}$, $(\tau,\xi) \in \mathbb{R}^{n+1}$, and we put $\dot{T}^*\tilde{X} = T^*\tilde{X} \setminus \tilde{X}$. If $i : X \longrightarrow \tilde{X}$ is the canonical immersion and $i^* : T^*_X\tilde{X} \longrightarrow T^*X$ is the associated map, we put $N^*X = (i^*)^{-1}(0)$, the conormal bundle of X in $\tilde{X}$.

If V is an open subset of $\tilde{X}$ (or X), by $L^m(V;p\times q)$, $m \in \mathbb{R}$, $p,q \in \{1,2,\dots\}$, we denote the space of $p\times q$ matrices $A = (A_{ij})$ of classical properly supported pseudo differential operators (p d o's) of order m defined on V (we omit $p\times q$ when $p = q = 1$). If E is a vector space, E^p, $p \in \{1,2,\dots\}$, denotes the product of p copies of E.

We consider now the system

$$Pu = t\,\partial_t I_N\, u - B(x,D_x)u = f \tag{1.1}$$

where $B \in L^0(X;N\times N)$ and $u,f \in \mathcal{D}'(\tilde{X})^N$ (I_N being the identity matrix of $\mathbb{C}^N$).

To give a meaning to (1.1) we suppose that the vector $u = (u_1,\dots,u_N)$ satisfies the assumption $WF(u) = \bigcup_{j=1}^{N} WF(u_j) \subset \dot{T}^*\tilde{X} \setminus N^*X$.

Let Σ_1 (resp. Σ_2) be the hypersurface of $\dot{T}^*\tilde{X}$ defined by $t = 0$ (resp. $\tau = 0$) and put $\Sigma_0 = \Sigma_1 \cap \Sigma_2$, $\Sigma = \Sigma_1 \cup \Sigma_2$,

$$\Sigma_1^{\pm} = \{(0,x,\tau,\xi) \in \Sigma_1 \mid \pm\tau > 0\},\ \Sigma_2^{\pm} = \{(t,x,0,\xi) \in \Sigma_2 \mid \pm t > 0\}.$$

It is well known that $WF(u) \setminus WF(Pu) \subset \Sigma$ and that the structure of $(WF(u) \setminus WF(Pu))\, \cap\, (\Sigma_j \setminus \Sigma_0)$, $j = 1,2$, if $WF(u) \cap N^*X = \emptyset$, is taken care of by the results of Duistermaat - Hörmander [10].

The situation is more complicated near Σ_o, where we expect that a phenomenon of branching of the singularities may appear.

To give a precise description we introduce some relations on $\Sigma\times\Sigma$. If $\rho_o =$ $= (t_o,x_o,\tau^{(o)},\xi^{(o)}) \in \Sigma_1$ (resp. Σ_2) define:

$$(1.2)\quad \begin{cases} \gamma_1^{\pm}(\rho_o) = \{(0,x_o,\tau,\xi^{(o)}) \in \Sigma_1 \mid \pm(\tau-\tau^{(o)}) \geq 0\} \\ \gamma_2^{\pm}(\rho_o) = \{(t,x_o,0,\xi^{(o)}) \in \Sigma_2 \mid \pm(t-t_o) \geq 0\} \end{cases}$$

We now define four relations Γ^{++}, Γ^{+-}, Γ^{-+}, Γ^{--} on $\Sigma\times\Sigma$ depending on the possible orientations of the bicharacteristics. Precisely, denoting by $\pi : \Sigma_j \longrightarrow \Sigma_o$, $j=1,2$, the canonical projection, we put :

$$\Gamma^{++} = \begin{cases} (\rho = (t,x,\tau,\xi),\rho' = (s,y,\sigma,\eta)) \in \Sigma\times\Sigma \mid \pi(\rho) = \pi(\rho') = \hat{\rho}, \\ \rho \in \gamma_1^{+}(\rho') \text{ if } \rho' \in \Sigma_1^{+},\ \rho \in \gamma_1^{+}(\rho') \cup \gamma_2^{+}(\hat{\rho}) \text{ if } \rho' \in \Sigma_1^{-} \cup \Sigma_o ; \\ \rho \in \gamma_2^{+}(\rho') \text{ if } \rho' \in \Sigma_2^{+},\ \rho \in \gamma_2^{+}(\rho') \cup \gamma_1^{+}(\hat{\rho}) \text{ if } \rho' \in \Sigma_2^{-} \cup \Sigma_o \end{cases}$$

(1.3)

$$\Gamma^{+-} = \begin{cases} (\rho = (t,x,\tau,\xi),\rho' = (s,y,\sigma,\eta)) \in \Sigma\times\Sigma \mid \pi(\rho) = \pi(\rho') = \hat{\rho}, \\ \rho \in \gamma_1^{+}(\rho') \text{ if } \rho' \in \Sigma_1^{+},\ \rho \in \gamma_1^{+}(\rho') \cup \gamma_2^{-}(\hat{\rho}) \text{ if } \rho' \in \Sigma_1^{-} \cup \Sigma_o ; \\ \rho \in \gamma_2^{-}(\rho') \text{ if } \rho' \in \Sigma_2^{-},\ \rho \in \gamma_2^{-}(\rho') \cup \gamma_1^{+}(\hat{\rho}) \text{ if } \rho' \in \Sigma_2^{+} \cup \Sigma_o \end{cases}$$

Γ^{-+} and Γ^{--} are defined as Γ^{+-}, Γ^{++} interchanging the roles of Σ_1 and Σ_2.

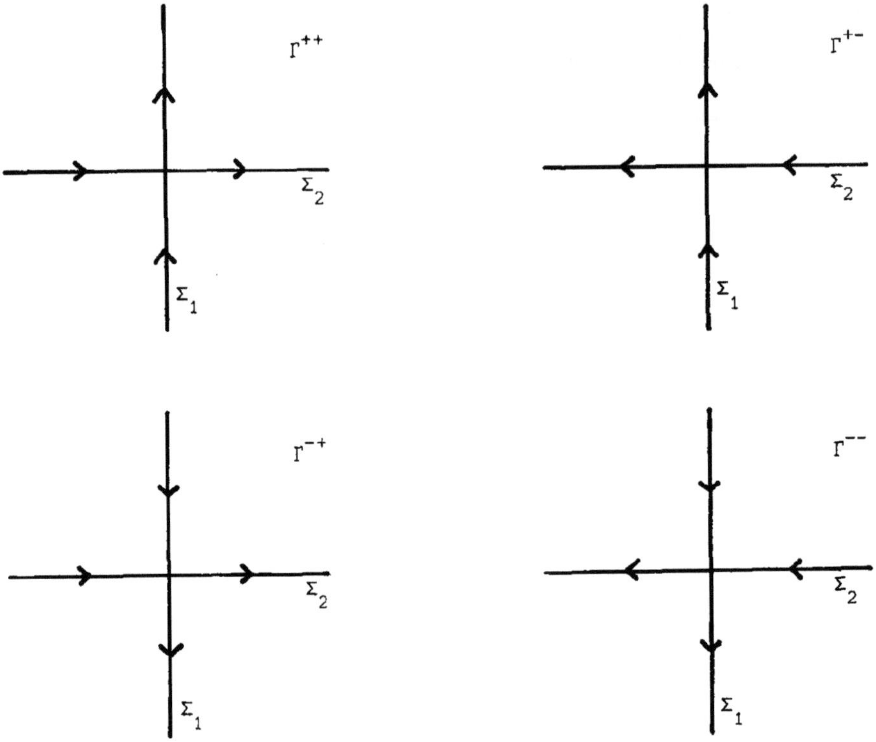

We have the following crucial result.

PROPOSITION 1.1. Let $\mathscr{R}$ be any one of the four relations $\Gamma^{\pm\pm}$ and let $\rho_o \in \Sigma_o$.

There exist a conic neighborhood V_{ρ_o} of ρ_o with $V_{\rho_o} \cap N^*X = \emptyset$ and a properly supported operator

$$E_{\mathscr{R}} : C_o^{\infty}(\tilde{X})^N \longrightarrow \mathscr{E}'(\tilde{X})^N$$

such that :

(i) $\quad WF'(E_{\mathscr{R}}) \cap (V_{\rho_0} \times T^*\tilde{X}) \subset (\Delta_{\dot{T}^*\tilde{X}} \cup \mathscr{R}) \cap (V_{\rho_0} \times T^*\tilde{X})$

(ii) $\quad WF'(E_{\mathscr{R}} P - I) \cap (V_{\rho_0} \times T^*\tilde{X}) = \emptyset$,

where $\Delta_{\dot{T}^*\tilde{X}}$ denotes the diagonal of $\dot{T}^*\tilde{X} \times \dot{T}^*\tilde{X}$.

The above result has been proved by N. Hanges [13, Proposition 3.6] in the scalar case $N = 1$. The same constructions can be carried over to the vector-valued situation; the necessary estimates for the symbols of the operators $E_{\mathscr{R}}$ are obtained by observing that for z in a region of $\mathbb{C}$ where $\log z$ is defined we have

$$z^{b_0(x,\xi)} = (2\pi i)^{-1} \int_\Gamma z^\zeta (\zeta - b_0(x,\xi))^{-1} d\zeta ,$$

where $b_0(x,\xi)$ is the principal symbol of $B(x,D_x)$ and Γ is a contour enclosing the spectrum of $b_0(x,\xi)$; we refer the reader to [13] for the details.

When $b_0(x,\xi)$ satisfies suitable conditions, one can construct microlocal left parametrices for P near Σ_0 having a "propagation relation" rather different from that of the operators $E_{\mathscr{R}}$.

Precisely, put:

$$(1.4) \quad \begin{cases} \Lambda = \{(\rho = (t,x,\tau,\xi), \rho' = (s,y,\sigma,\eta)) \in \Sigma \times \Sigma \mid \pi(\rho) = \pi(\rho'), \frac{\tau}{\sigma} \in [0,1], \frac{s}{t} \in [0,1]\} \\ \Lambda' = \{(\rho = (t,x,\tau,\xi), \rho' = (s,y,\sigma,\eta)) \in \Sigma \times \Sigma \mid \pi(\rho) = \pi(\rho'), \frac{\sigma}{\tau} \in [0,1], \frac{t}{s} \in [0,1]\}, \\ \text{we say } \frac{\lambda}{0} \in [0,1] \text{ iff } \lambda = 0 \end{cases}$$

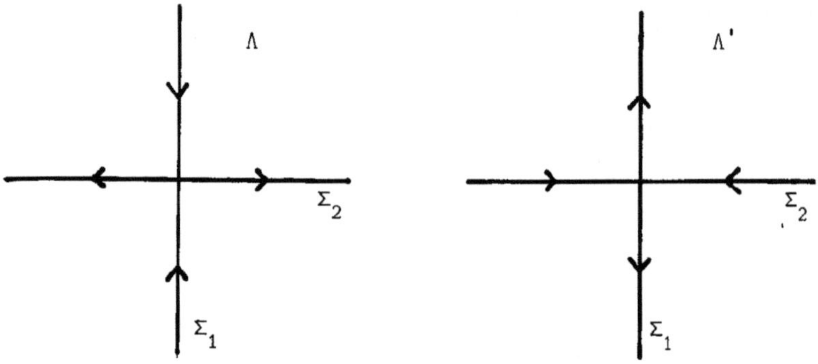

We have the following result.

PROPOSITION 1.2. Let $\rho_o \in \Sigma_o$ and denote by $\sigma(b_o(x,\xi))$ the spectrum of the principal symbol of $B(x,D_x)$. Then:

(a) if $\sigma(b_o(\rho_o)) \cap \{0,1,2,\dots\} = \emptyset$, there exist a conic neighborhood V_{ρ_o} of ρ_o with $V_{\rho_o} \cap N^*X = \emptyset$ and a properly supported operator

$$E_\Lambda \; : \; C_o^\infty(\tilde{X})^N \longrightarrow \mathscr{E}'(\tilde{X})^N$$

such that:

(i) $WF'(E_\Lambda) \cap (V_{\rho_o} \times T^*\tilde{X}) \subset (\Delta_{\dot{T}^*\tilde{X}} \cup \Lambda) \cap (V_{\rho_o} \times T^*\tilde{X})$.

(ii) $WF'(E_\Lambda P - I) \cap (V_{\rho_o} \times T^*\tilde{X}) = \emptyset$.

(b) If $\sigma(b_o(\rho_o)) \cap \{-1, -2,\dots\} = \emptyset$, there exist a conic neighborhood V_{ρ_o} of ρ_o with $V_{\rho_o} \cap N^*X = \emptyset$ and a properly supported operator

$$E_{\Lambda'} : C_0^\infty(\tilde{X})^N \longrightarrow \mathscr{D}'(\tilde{X})^N$$

such that

(i) $WF'(E_{\Lambda'}) \cap (V_{\rho_o} \times T^*\tilde{X}) \subset (\Delta_{\dot{T}^*\tilde{X}} \cup \Lambda') \cap (V_{\rho_o} \times T^*\tilde{X})$

(ii) $WF'(E_{\Lambda'} P - I) \cap (V_{\rho_o} \times T^*\tilde{X}) = \emptyset$.

The above result was proved by N. Hanges [13, Proposition 4.4] in the scalar case $N = 1$; the extension to the vector-valued case can be carried over following the same arguments of [13] and taking into account the remark after Proposition 1.1.

As a trivial consequence of Propositions 1.1, 1.2 the following result holds:

THEOREM 1.1. Let $u \in \mathcal{D}'(\tilde{X})^N$ with $WF(u) \cap N^*X = \emptyset$ and let $\rho_o \in \Sigma_o \setminus WF(Pu)$.

Then:

(1) If for every $j \in \{1,2\}$ there is a choice of the sign + or - for which $(\gamma_j^{\pm}(\rho_o) \setminus \{\rho_o\}) \cap WF(u) = \emptyset$, $j = 1,2$, then $\rho_o \notin WF(u)$.

(2) If $\sigma(b_o(\rho_o)) \cap \{0,1,2,\dots\} = \emptyset$ and

$$[(\gamma_1^+(\rho_o) \cup \gamma_1^-(\rho_o)) \setminus \{\rho_o\}] \cap WF(u) = \emptyset , \quad \text{then} \quad \rho_o \notin WF(u).$$

(3) If $\sigma(b_o(\rho_o)) \cap \{-1, -2, \dots\} = \emptyset$ and

$$[(\gamma_2^+(\rho_o) \cup \gamma_2^-(\rho_o)) \setminus \{\rho_o\}] \cap WF(u) = \emptyset , \quad \text{then} \quad \rho_o \notin WF(u).$$

For $N=1$, Theorem 1.1 was proved by N. Hanges [13] (see also, for a less constructive proof V. Ya. Ivrii [16], R. Melrose [22] and for a slightly different case S. Alinhac [3]); for similar results in the hyperfunction setting see T. Miwa [24], M. Kashiwara, T. Kawai & T. Oshima [18], H. Tahara [28], T. Ôaku [26].

The results of Theorem 1.1 are sharp. Under suitable conditions on the full symbol $b(x,\xi)$ of the operator $B(x,D_x)$, there is a distribution $u \in \mathcal{D}'(\tilde{X})^N$ such that $\rho_0 \notin WF(Pu)$ and $WF(u)$ contains three half-bicharacteristics; for a detailed description in the scalar case see Ivrii [16, Theorem 4.11].

For the applications we have in mind we now give a result on the microlocal structure of u near ρ_0 when $\sigma(b_0(\rho_0)) \cap Z \neq \emptyset$.

THEOREM 1.2. Let k be a non-negative integer and $\rho_0 \in \Sigma_0 \setminus WF(Pu)$. Then:

(2)' if $\sigma(b_0(\rho_0)) \cap \{k+1, k+2, \ldots\} = \emptyset$ and

$[(\gamma_1^+(\rho_0) \cup \gamma_1^-(\rho_0)) \setminus \{\rho_0\}] \cap WF(u) = \emptyset$, there exist distributions $v_j \in \mathcal{D}'(X)^N$, $j=0,1,\ldots,k$ such that

$$\rho_o \notin WF(u - \sum_{j=0}^{k} t^j \otimes v_j(x)) . \tag{1.5}$$

(3)' If $k \geq 1$, $\sigma(b_0(\rho_0)) \cap \{-k-1, -k-2, \ldots\} = \emptyset$ and

$[(\gamma_2^+(\rho_0) \cup \gamma_2^-(\rho_0)) \setminus \{\rho_0\}] \cap WF(u) = \emptyset$, there exist distributions $w_j \in \mathcal{D}'(X)^N$, $j=0,1,\ldots,k-1$, such that

$$\rho_o \notin WF(u - \sum_{j=0}^{k-1} \delta_t^{(j)} \otimes w_j(x)), \quad \delta_t^{(j)} = \partial_t^j \delta_t . \tag{1.6}$$

Proof of (2)': Since, by hypothesis, $WF(u) \cap N^*X = \emptyset$, u and $Pu=f$ have traces of all order at $t=0$. Put $\varphi_j = \partial_t^j u\,|_{t=0}$, $\psi_j = \partial_t^j f\,|_{t=0}$, $j=0,1,\ldots;$

from (1.1) we obtain

$$(1.7) \qquad (j\, I_N - B(x,D_x))\varphi_j(x) = \psi_j(x) , \qquad j = 0,1,2,\dots$$

Writing $v = u - \sum_{j=0}^{k} \frac{t^j}{j!} \otimes \varphi_j(x)$, $g = f - \sum_{j=0}^{k} \frac{t^j}{j!} \otimes \psi_j(x)$

and taking into account (1.7) we obtain

$$(1.8) \qquad P v = g$$

Let $\mathbb{R}_{k+1}$ be the $(k+1)^{th}$ order Taylor remainder operator

$$(1.9) \qquad \mathbb{R}_{k+1}\, w(t,x) = \frac{1}{(k+1)!} \int_0^1 (1-\sigma)^k \, (\partial_t^{k+1} w)(t\sigma,x)\, d\sigma .$$

It can be verified (Cf. Hanges [13]) that :

$$(1.10) \qquad WF'(\mathbb{R}_{k+1}) \subset \Delta_{\dot{T}^*\tilde{X}} \cup \Lambda \cup \mathcal{F}$$

where Λ is defined in (1.4) and

$$(1.10)' \qquad \begin{aligned} \mathcal{F} = & \{((t,x,\tau,0),\ (0,y,\sigma,0)) \mid t\tau = 0,\ \tfrac{\tau}{\sigma} \in [0,1]\} \\ & \cup \{((t,x,\tau,0),\ (t,y,\tau,0)) \mid \tau \neq 0\} . \end{aligned}$$

It follows from (1.10), (1.10)' that $\mathbb{R}_{k+1}$ can be extended to distributions $w \in \mathcal{E}'(X)^N$ such that $WF(w) \cap N^*X = \emptyset$.

We can obviously suppose that $u \in \mathcal{E}'(\tilde{X})^N$ so that we can write

$v = t^{k+1}\,\mathbb{R}_{k+1}(v)$, $g = t^{k+1}\,\mathbb{R}_{k+1}(g)$. Putting $w = \mathbb{R}_{k+1}(v)$, $h = \mathbb{R}_{k+1}(g)$, one obtains $t\,\partial_t(t^{k+1} w) = (k+1)t^{k+1} w + t^{k+1}\, t\,\partial_t w = t^{k+1} Bw + t^{k+1} h$, i.e.

$$(1.11) \qquad t^{k+1}\,[t\,\partial_t w - (B(x,D_x) - (k+1)I_N)\,w - h] = 0 .$$

The following facts are easily established :

$$(1.12) \qquad \begin{aligned} &\text{i)} \quad (WF(w) \cup WF(h)) \cap N^*X = \emptyset . \\ &\text{ii)} \quad \rho_0 \notin WF(h),\ [(\gamma_1^+(\rho_0) \cup \gamma_1^-(\rho_0)) \setminus \{\rho_0\}] \cap WF(w) = \emptyset . \end{aligned}$$

From eq. (1.11) it follows that the distribution $t\,\partial_t w - (B-(k+1)I)w - h$ is supported in $\{0\} \times X$ and, since its wave front set is disjoint from N^*X, it can be easily recognized that

$$(1.13) \qquad t\,\partial_t w = (B(x,D_x) - (k+1)I_N)w + h.$$

From (1.12) ii) and Theorem 1.1 (2) the thesis follows, since

$$\sigma(b_0(\rho_0) - (k+1)I_N) \cap \{0,1,\dots\} = \emptyset .$$

<u>Proof of (3)'</u>. As above we suppose that $u \in \mathscr{E}'(\tilde{X})^N$. The idea is to express u, for $t \neq 0$, as $u = \partial_t^k v$ for some distribution $v \in \mathscr{E}'(\mathbb{R}_t \times X)^N$. For this define

$$(1.14) \qquad v = \mathscr{P}_k u + \mathscr{Q}_k u ,$$

where $\mathscr{P}_k$, $\mathscr{Q}_k$ are $N \times N$ diagonal matrices of operators with the following distribution kernels :

$$(1.15) \quad \begin{cases} p_k(t,x;s,y) = \dfrac{(-1)^k}{(k-1)!} Y(t)(s-t)_+^{k-1} \otimes \delta(x-y) , \\ \\ q_k(t,x;s,y) = \dfrac{1}{(k-1)!} Y(-t)(t-s)_+^{k-1} \otimes \delta(x-y) , \end{cases}$$

$Y(t)$ being the Heaviside function.

A computation of $WF'(\mathscr{P}_k)$, $WF'(\mathscr{Q}_k)$ shows that v is well defined (since $WF(u) \cap N^*X = \emptyset$) and that

$$[(\gamma_2^+(\rho_o) \cup \gamma_2^-(\rho_o)) \smallsetminus \{\rho_o\}] \cap WF(v) = \emptyset .$$

For $t \neq 0$, formula (1.14) reads :

$$(1.14)' \quad v(t,x) = \begin{cases} \dfrac{(-1)^k}{(k-1)!} \displaystyle\int_t^{+\infty} (s-t)^{k-1} u(s,x)ds, & t > 0 \\ \\ \dfrac{1}{(k-1)!} \displaystyle\int_{-\infty}^{t} (t-s)^{k-1} u(s,x)ds, & t < 0 \end{cases}$$

Moreover, $\partial_t^k v = u$ for $t \neq 0$.

A computation shows that

$$(1.16) \qquad u = \partial_t^k v + \sum_{j=0}^{k-1} \frac{(-1)^j}{j!} \delta_t^{(j)} \otimes \Theta_j(x) ,$$

where $\Theta_j(x) = \int_{-\infty}^{+\infty} s^j u(s,x)ds$ is the push-forward of $t^j u$ under the projection of $\mathbb{R} \times X$ onto X. Write (1.16) as $u = \partial_t^k v + \sum_{j=0}^{k-1} \delta_t^{(j)} \otimes w_j(x)$ and observe that

$$(1.17) \qquad (t\partial_t I_N - B(x,D_x)) [\partial_t^k v + \sum_{j=0}^{k-1} \delta_t^{(j)} \otimes w_j(x)] = f .$$

Since $t\,\partial_t \delta_t^{(j)} = -(j+1)\,\delta_t^{(j)}$, multiplying (1.17) by t^k and taking into account that $t^k \delta_t^{(j)} = 0$ for $j < k$, we obtain

$$(1.18) \qquad t^k (t\partial_t I_N - B(x,D_x)) (\partial_t^k v) = t^k f.$$

Using the relations

$$(1.19) \qquad \begin{cases} t^k t\,\partial_t = (t\,\partial_t - k) t^k , \\ t^k \partial_t^k = t\,\partial_t (t\,\partial_t - 1) \ldots (t\,\partial_t - k + 1) , \end{cases}$$

we have that

(1.19)' $$(t\,\partial_t I_N - B(x,D_x))\ [t\,\partial_t((t\,\partial_t - 1)I_N)\ \dots\ ((t\,\partial_t - k+1)I_N)]\ v = t^k f\ .$$

Define

$$\varphi_1 = v\ ,\ \varphi_2 = (t\,\partial_t - k+1)\varphi_1\ ,\ \dots\ ,\ \varphi_{k+1} = t\,\partial_t\,\varphi_k .$$

From (1.19)' we obtain the system

(1.20) $$\begin{cases} t\,\partial_t\varphi_1 = (k-1)\varphi_1 + \varphi_2 \\ t\,\partial_t\varphi_2 = (k-2)\varphi_2 + \varphi_3 \\ \dots\dots\dots\dots\dots\dots \\ t\,\partial_t\varphi_k = \varphi_{k+1} \\ t\,\partial_t\varphi_{k+1} = (B(x,D_x) + k\,I_N)\varphi_{k+1} + t^k f\ . \end{cases}$$

Thus the big vector $\varphi = (\varphi_1, \dots, \varphi_{k+1})$ satisfies the system $(t\,\partial_t I_{N(k+1)} - \mathscr{B}(x,D_x))\varphi = g$, where $g = (0,0,\dots,t^k f)$ and $\mathscr{B} \in L^0(X; N(k+1) \times N(k+1))$ with a principal symbol whose spectrum at ρ_0 is given by

(1.21) $$\sigma(b_0(\rho_0) + k\,I_N)\ \cup\ \{0,1,\dots,k-1\}\ .$$

Therefore, by Theorem 1.1 (3), we obtain that $\rho_0 \notin WF(\varphi)$ and thus (1.6) is proved.

q.e.d.

<u>Remarks.</u> Since $WF(\sum_{j=o}^{k} t^j \otimes v_j(x)) \subset \Sigma_2$ and $WF(\sum_{j=o}^{k-1} \delta_t^{(j)} \otimes w_j(x)) \subset \Sigma_1$,

we have an information on the singularities of u near ρ_o when $\sigma(b_o(\rho_o)) \cap Z \neq \emptyset$. Moreover, the v_j's are traces of u at $t=0$ while the w_j's are some momenta of u. We can thus state the following corollary.

COROLLARY.

(2)" Let the hypotheses of Theorem 1.2 (2)' be satisfied and suppose that $\sigma(b_o(\rho_o)) \cap \{0,1,\dots,k\} = \{h_1,\dots,h_r\}$, $0 \leq h_1 < \dots < h_r \leq k$, $r \geq 1$. Then there exist distributions $\varphi_1,\dots,\varphi_r \in \mathcal{D}'(X)^N$ such that

$$\rho_o \notin WF(u - \sum_{j=1}^{r} t^{h_j} \otimes \varphi_j(x)) . \tag{1.22}$$

Moreover, if $B(x,D_x) - h_j I_N$ is microlocally hypoelliptic at ρ_o, then $\rho_o \notin WF(t^{h_j} \otimes \varphi_j(s))$.

(3)" Let the hypotheses of Theorem 1.2 (3)' be satisfied and suppose that $\sigma(b_o(\rho_o)) \cap \{-1,-2,\dots,-k\} = \{-\ell_1,\dots,-\ell_s\}$, $1 \leq \ell_1 < \dots < \ell_s \leq k$, $s \geq 1$. Then there exist distributions $\psi_1,\dots,\psi_s \in \mathcal{D}'(X)^N$ such that

$$\rho_o \notin WF(u - \sum_{j=1}^{s} \delta_t^{(\ell_j - 1)} \otimes \psi_j(x)) \tag{1.23}$$

Moreover, if $B(x,D_x) + \ell_j I_N$ is microlocally hypoelliptic at ρ_o, then $\rho_o \notin WF(\delta_t^{(\ell_j - 1)} \otimes \psi_j(x))$.

Proof of (2)". From (1.5) we have $\rho_o = (0,x_o,0,\xi^{(o)}) \notin WF(u - \sum_{j=0}^{k} t^j \otimes v_j(x))$.

Since $(\{\rho_o\} \cup N^*X) \cap WF(f) = \emptyset$, we have that $(x_o,\xi^{(o)}) \notin WF(\partial_t^j f_{|t=0})$, $j \geq 0$.

The ellipticity of $j\, I_N - B(x,D_x)$ in ρ_o when $j \notin \{h_1,\dots,h_r\}$ and relation (1.7) yield $(x_o,\xi^{(o)}) \notin WF(v_j)$ for $j \notin \{h_1,\dots,h_r\}$ and thus (1.22) is proved.

The microhypoellipticity of $B(x,D_x) - h_j I_N$ means that $\rho_o \notin WF(t^j \otimes v_j)$ if $\rho_o \notin WF(t^j \otimes (B(x,D_x) - h_j I_N) v_j)$ and thus the conclusion follows.

Proof of (3)". From the relation $t\,\partial_t I_N u - Bu = f$ we obtain, supposing $u \in \mathscr{E}'(\tilde{X})^N$:

$$(1.24)\qquad \begin{aligned} &\int_{-\infty}^{+\infty} t^j[t\partial_t u - Bu]dt = \int_{-\infty}^{+\infty} t\partial_t(t^j u)dt \\ &-(j\, I_N + B)\int_{-\infty}^{+\infty} t^j u(t)dt = \int_{-\infty}^{+\infty} t^j f(t)dt\ . \end{aligned}$$

Thus

$$(1.25)\qquad -((j+1)I_N + B(x,D_x))\int_{-\infty}^{+\infty} t^j u(t)dt = \int_{-\infty}^{+\infty} t^j f(t)dt\ ,\quad j \geq 0.$$

Since $WF(\int_{-\infty}^{+\infty} t^j f(t)dt) \subset \{(x,\xi) \in \dot{T}^*X \mid \exists\, t, (t,x,0,\xi) \in WF(f)\}$ and since $(\gamma_2^+(\rho_o) \cup \gamma_2^-(\rho_o)) \cap WF(f) = \emptyset$, we obtain that $(x_o,\xi^{(o)}) \notin WF(\int_{-\infty}^{+\infty} t^j f(t)dt)$.

From (1.6) we have $\rho_o \notin WF(u - \sum_{j=0}^{k-1} \delta_t^{(j)} \otimes w_j(x))$. The ellipticity of $(j+1)I_N + B(x,D_x)$ in ρ_o when $-(j+1) \notin \{-\ell_1, \dots, -\ell_s\}$ and relation (1.25) yield $(x_o,\xi^{(o)}) \notin WF(w_j)$ for $(j+1) \notin \{\ell_1, \dots, \ell_s\}$ and thus (1.23) is proved.

The last conclusion follows in the same way as above.

q.e.d.

2. GENERAL FUCHSIAN SYSTEMS

In this Section we consider a Fuchsian system

$$(2.1)\qquad Pu = t\partial_t I_N u - A(t,x,D_t,D_x)u = f ,$$

where $A \in L^0(\tilde{X}; N\times N)$ and $u,f \in \mathcal{D}'(\tilde{X})^N$; we denote by $a(t,x;\tau,\xi)$ the full (matrix) symbol of the operator A and write its asymptotic expansion as $a(t,x;\tau,\xi) \sim a_0(t,x;\tau,\xi) + a_{-1}(t,x;\tau,\xi) + \ldots$. With the same notation of Section 1, we are interested in studying $WF(u) \setminus WF(Pu)$ near a point $\rho_0 = (0,x_0;0,\xi^{(0)}) \in \Sigma_0 = \Sigma_1 \cap \Sigma_2$. The natural idea is to try to find a p d o $Q \in L^0(\tilde{X};N\times N)$, elliptic near ρ_0, such that $Q^{-1}PQ \equiv t\partial_t I_N - B(x,D_x)$ (near ρ_0) for some operator $B \in L^0(X;N\times N)$ and then apply the results of Section 1. It turns out that in the scalar case $N=1$ such an intertwining operator Q does exist (see N. Hanges [12, Lemmas 2.4, 2.5]), but in the vector valued situation we find an obstruction which is related to a classical condition encountered in the study of ordinary differential equations of Fuchs type (see e.g. B. Malgrange [21]). Precisely it will be shown that for $N > 1$ an intertwining operator Q with the above properties exists if the following *Fuchs condition* is satisfied at the point $\rho_0 \in \Sigma_0$:

$$(F)_{\rho_0}\qquad \sigma(a_0(\rho_0) + j\, I_N) \cap \sigma(a_0(\rho_0)) = \emptyset\ ,\ \forall j \in \mathbb{Z}\setminus\{0\} ,$$

where $\sigma(\,\cdot\,)$ denotes the spectrum and a_0 is the (matrix) principal symbol of the operator $A(t,x,D_t,D_x)$. Note that $(F)_{\rho_0}$ is satisfied if $N=1$. If condition $(F)_{\rho_0}$ is not satisfied, we use an idea of M. Kashiwara - T. Oshima [19] (see also the recent results by T. Ōaku [26]) to construct, starting from the original system

(2.1) , a new system $\mathscr{P}U = (t\,\partial_t I_M - \mathscr{B}(x,D_x))U = F$, with $M \gg N$, $WF(F) = WF(f)$, $WF(U) = WF(u)$ near ρ_o, moreover, the spectrum of the principal symbol of the operator $\mathscr{B} \in L^o(X;M\times M)$ at ρ_o is equal to $\sigma(a_o(\rho_o))$. Unfortunatèly this construction is quite long and we prepare it starting with a series of Lemmas.

LEMMA 2.0 . Let M be a C^∞ countable manifold and suppose we are given a function $F \in C^\infty(R^{h+k}_{(t,s)} \times M;\ \mathscr{L}(C^d,C^d))$ and the differential operator $\mathscr{L} = \sum_{j=1}^{h} \lambda_j t_j \partial_{t_j} - \sum_{i=1}^{k} \mu_i s_i \partial_{s_i}$, where the λ_j, μ_i are fixed positive real numbers.

Denote by $C^\infty_{flat}(\mathbb{R}^{h+k}\times M)^d$ the space of all (vector-valued) functions $f \in C^\infty(\mathbb{R}^{h+k}\times M)^d$ for which $\partial^\alpha_t f\,\big|_{t=0} = \partial^\beta_s f\,\big|_{s=0} = 0$ for every α,β .

Then, for every $T > 0$ and for every $g \in C^\infty_{flat}(\mathbb{R}^{h+k}\times M)^d$, there exists a function $f \in C^\infty_{flat}(\mathbb{R}^{h+k}\times M)^d$ such that

(2.2) $$(\mathscr{L} - F)f = g \quad \text{on} \quad C_T = \{(t,s)\,|\ |t_j| < T,\ |s_i| < T,\ \forall j,i\}\times M .$$

<u>Proof</u>. Using a smooth cut-off function we can suppose that $F = g = 0$ out of the strip C_{2T} defined in $\mathbb{R}^{h+k}\times M$ by the condition $|s_i| \leq 2T$, $i = 1,\dots,k$. Consider now the operator

(2.3) $$Rg(t,s;z) = \int_0^1 g(\rho^\lambda t,\rho^{-\mu}s;z)\,\frac{d\rho}{\rho} \quad , \quad z \in M ,$$

where $\rho^{\lambda} t = (\rho^{\lambda_1} t_1,\dots,\rho^{\lambda_h} t_h)$, $\rho^{-\mu} s = (\rho^{-\mu_1} s_1,\dots,\rho^{-\mu_k} s_k)$.

It is easily seen that Rg is continuous and vanishes out of the same strip and $\mathscr{L}\,Rg = g$.

For every $N,M \geq 0$ and for every seminorm q of $C^{\infty}(M)$ define the following seminorms on $C^{\infty}_{flat}(\mathbb{R}^{h+k} \times M)^d$:

$$(2.4) \qquad p_{N,M,q}(f) = \sum_{\langle\alpha,\lambda\rangle+\langle\beta,\mu\rangle \leq M} \sup_{C_{2T}} q\left[\prod_{j=1}^{h} |t_j|^{-\frac{(N+M)}{\lambda_j}} \prod_{i=1}^{k} |s_i|^{-\frac{M}{\mu_i}} \partial_t^{\alpha} \partial_s^{\beta} f\right]$$

Taking N large with respect to M it is easily seen that $p_{N,M,q}(f) < \infty$, so that $Rg \in C^{\infty}(C_{2T} \times M)^d$ and Rg is flat at $t=0$ or $s=0$.

To construct the required f we use Picard's method by defining $\varphi_0 = 0$, $\varphi_{n+1} = R(F\varphi_n + g)$, $n = 0,1,\dots$.

Since for every $M \geq 0$ and for every seminorm q we can find $N = N(M,q)$ such that $p_{N(M,q),M,q}(\varphi_{n+1} - \varphi_n) \leq 1/2 \; p_{N(M,q),M,q}(\varphi_n - \varphi_{n-1})$ for every $n \geq 1$, we conclude that φ_n converges to a function $\varphi \in C^{\infty}_{flat}(C_{2T} \times M)^d$ which, by construction, solves the system $(\mathscr{L} - F)\varphi = g$. Modifying φ out of $C_T \times M$ we obtain a function $f \in C^{\infty}_{flat}(\mathbb{R}^{h+k} \times M)^d$ for which (2.2) holds.

q.e.d.

LEMMA 2.1. Let M be a C^{∞} countable manifold and let $a(z;t,\tau)$ be a smooth $N \times N$ matrix-valued function defined for $(z;t,\tau) \in M \times \mathbb{R}^2$. Assume that:

1) $a(z;t,\tau) = (a_{ij}(z;t,\tau))_{i,j=1,\dots,k}$, where the a_{ij} are $N_i \times N_j$ matrices, $N_1+\dots+N_k = N$, $k \geq 2$.

2) At a point $z_o \in M$ we have:

i) $a_{ij}(z_o;0,0) = 0$, for $i \neq j$

ii) Putting $a_{ii}(z_o;0,0) = \Lambda_i$,

$$\sigma(\Lambda_i + r\, I_{N_i}) \cap \sigma(\Lambda_j) = \emptyset \quad \begin{cases} i,j = 1,\dots,k,\ i \neq j \\ r \in \mathbb{Z} \end{cases}$$

Then there is a neighborhood $V \times (-T,T) \times (-T,T)$ of $(z_o;0,0)$ and two smooth $N \times N$ matrix-valued functions $q(z;t;\tau)$, $\tilde{a}(z;t,\tau)$, defined on $M \times \mathbb{R}^2$, such that

I) $q(z_o;0,0) = I_N$.

II) $\tilde{a}(z;t,\tau)$ is in diagonal block form with blocks of dimension $N_i \times N_i$, $i = 1,\dots,k$, and $\tilde{a}(z_o;0,0) = a(z_o;0,0)$.

III) $Lq = (t\,\partial_t - \tau\,\partial_\tau)q - \{aq - q\tilde{a}\} = 0$ on $V \times (-T,T) \times (-T,T)$.

<u>Proof.</u> Without loss of generality assume $k = 2$; the general case will follow by induction on k. Consider the Taylor expansion $a(z;t,\tau) \sim \sum a^{rs}(z)t^r \tau^s$ of a and seek for formal power series $\sum q^{rs}(z)t^r \tau^s$, $\sum \tilde{a}^{rs}(z)t^r \tau^s$ such that $(Lq)^{rs}(z) = 0$ for every r,s. We have :

$$(Lq)^{rs}(z) = (r-s)q^{rs}(z) \tag{2.5}$$
$$- \sum_{r'=0}^{r} \sum_{s'=0}^{s} \{a^{r-r',s-s'}(z)q^{r's'}(z) - q^{r's'}(z)\tilde{a}^{r-r',s-s'}(z)\}.$$

We wish to choose q^{rs} and a^{rs} so that $q^{00}(z_o) = I_N$, $\tilde{a}^{00}(z_o) = a(z_o;0,0)$ and $(Lq)^{rs}(z) = 0$ for every r,s.

First consider the case $r = s = 0$; the matrix-valued functions $q^{00}(z),\tilde{a}^{00}(z)$

must satisfy the equation

$$(2.6) \qquad (Lq)^{oo}(z) = -\{a^{oo}(z)q^{oo}(z) - q^{oo}(z)\tilde{a}^{oo}(z)\} = 0.$$

Let $(\hat{q},\tilde{a}) = \begin{pmatrix} \tilde{a}_{11} & q_{12} \\ q_{21} & \tilde{a}_{22} \end{pmatrix}$ and $(\hat{q},I_N) = \begin{pmatrix} I_{N_1} & q_{12} \\ q_{21} & I_{N_2} \end{pmatrix}$,

$(0,\tilde{a}) = \begin{pmatrix} \tilde{a}_{11} & 0 \\ 0 & \tilde{a}_{22} \end{pmatrix}$.

Consider the quadratic expression

$$(2.7) \qquad (\hat{q},\tilde{a}) \longmapsto F(z;\hat{q},\tilde{a}) = a^{oo}(z)(\hat{q},I_N) - (\hat{q},I_N)(0,\tilde{a}).$$

When $z = z_o$, $\tilde{a} = a(z_o;0,0)$, $\hat{q} = 0$ the derivative $dF_{(\hat{q},\tilde{a})}$ is given by :

$$(\Delta\hat{q},\ \Delta\tilde{a}) \longmapsto dF_{(\hat{q},\tilde{a})}(\Delta\hat{q},\ \Delta\tilde{a}) =$$

$$(2.8) \qquad = \begin{pmatrix} -\Delta\tilde{a}_{11} & a^{oo}_{11}(z_o)\,\Delta q_{12} - \Delta q_{12}\,a^{oo}_{22}(z_o) \\ a^{oo}_{22}(z_o)\,\Delta q_{21} - \Delta q_{21}\,a^{oo}_{11}(z_o) & -\Delta\tilde{a}_{22} \end{pmatrix}$$

At this point we use the well known fact that if A(resp. B) is a $p \times p$ (resp. $q \times q$) matrix, then the equation $AQ - QB = R$ has a unique solution Q for any given matrix R ($\dim Q = \dim R = p \times q$) if (and only if) A and B have disjoint spectrum. Since, by hypothesis 2) ii), $\sigma(a^{oo}_{11}(z_o)) \cap \sigma(a^{oo}_{22}(z_o)) = \emptyset$, the derivative $dF_{(\hat{q},\tilde{a})}$ is surjective; therefore, by the implicit function theorem, there exists a neighborhood W of z_o on which smooth functions $(\hat{q}(z), \tilde{a}(z))$ are defined and verify $F(z; \hat{q}(z), \tilde{a}(z)) = 0$ on W, $\hat{q}(z_o) = 0$, $\tilde{a}(z_o) = a^{oo}(z_o)$. Letting

$$q^{oo}(z) = (\hat{q}(z), I_N) = \begin{bmatrix} I_{N_1} & q_{12}(z) \\ q_{21}(z) & I_{N_2} \end{bmatrix}, \quad \tilde{a}^{oo}(z) = \begin{bmatrix} \tilde{a}_{11}(z) & 0 \\ 0 & \tilde{a}_{22}(z) \end{bmatrix},$$

eq. (2.6) is satisfied on W.

We now use induction to determine $q^{rs}(z)$, $\tilde{a}^{rs}(z)$ so that $(Lq)^{rs}(z) = 0$ for every r,s. Suppose that we have fixed a neighborhood $W' \subset W$ of z_o so that for all $r' \leq r$, $s' \leq s$ with $(r',s') \neq (r,s)$, $q^{r's'}(z)$, $\tilde{a}^{r's'}(z) \in C^\infty(W')$ have been found satisfying $(Lq)^{r's'}(z) = 0$ on W'. It follows from (2.5) that q^{rs} and $\tilde{a}^{rs}$ must satisfy the equation:

$$(Lq)^{rs}(z) = (r-s)q^{rs}(z) \tag{2.9}$$
$$- \{a^{oo}(z)q^{rs}(z) - q^{rs}(z)\tilde{a}^{oo}(z)\} + q^{oo}(z)\tilde{a}^{rs}(z) = \text{known data} \in C^\infty(W').$$

Again let $(\hat{q},\tilde{a}) = \begin{bmatrix} \tilde{a}_{11} & q_{12} \\ q_{21} & \tilde{a}_{22} \end{bmatrix}$ and seek for a solution of (2.9) with q of the form $q = (\hat{q},0)$. Consider the linear operator

$$(\hat{q},\tilde{a}) \longmapsto F_{r-s}(z;\hat{q},\tilde{a}) = (r-s)(\hat{q},0) \tag{2.10}$$
$$- \{a^{oo}(z)(\hat{q},0) - (\hat{q},0)\tilde{a}^{oo}(z)\} + q^{oo}(z)(0,\tilde{a}).$$

When $z = z_0$,

$$(2.11)\quad F_{r-s}(z_0;\hat{q},\tilde{a}) = \begin{pmatrix} \tilde{a}_{11} & q_{12}\, a_{22}(z_0) - [a_{11}(z_0) - (r-s)I_{N_1}]\, q_{12} \\ q_{21}\, q_{11}(z_0) - [a_{22}(z_0) - (r-s)I_{N_2}]\, q_{21} & \tilde{a}_{22} \end{pmatrix}$$

Thanks to hypothesis 2) ii) we can take a fixed neighborhood $W' \subset W$ of z_0 such that $F_{r-s}(z;\cdot,\cdot)$ is smoothly invertible on W' . Thus $(\hat{q}(z),\ \tilde{a}(z)) \in C^\infty(W')$ are obtained and we define $q^{rs}(z) = \begin{pmatrix} 0 & q_{12}(z) \\ q_{21}(z) & 0 \end{pmatrix}$,

$\tilde{a}^{rs}(z) = \begin{pmatrix} \tilde{a}_{11}(z) & 0 \\ 0 & \tilde{a}_{22}(z) \end{pmatrix}$.

In conclusion we have constructed formal power series $\sum q^{rs}(z)t^r \tau^s$, $\sum \tilde{a}^{rs}(z)t^r \tau^s$ with q^{rs}, $\tilde{a}^{rs} \in C^\infty(W')$ such that $(Lq)^{rs}(z) = 0$ on W' for every r,s. Modifying the $q^{rs}(z)$, $\tilde{a}^{rs}(z)$ out of a neighborhood $V' \subset W'$ of z_0 and using Borel's Lemma, we can find two smooth matrix-valued functions $q(z,t,\tau)$, $\tilde{a}(z;t,\tau)$ defined on $\mathbb{R}^2 \times M$, satisfying conditions I), II) and such that $Lq = (t\,\partial_t - \tau\,\partial_\tau)q - \{aq - q\tilde{a}\} = g \in C^\infty_{flat}(\mathbb{R}^2 \times V')^{N\times N}$. Using Lemma 2.0 we find $q_1 \in C^\infty_{flat}(\mathbb{R}^2 \times V')^{N\times N}$ such that $Lq_1 = -g$ on $V' \times (-T,T) \times (-T,T)$. Modifying q_1 out of a neighborhood $V \subset V'$ of z_0 , we can suppose that $q_1 \in C^\infty_{flat}(\mathbb{R}^2 \times M)^{N\times N}$. Defining $q =$ the preceding $q + q_1$, we have that $q,\tilde{a} \in C^\infty(\mathbb{R}^2 \times M)^{N\times N}$ satisfy I), II) and that $Lq = (t\,\partial_t - \tau\,\partial_\tau)q - \{aq - q\tilde{a}\} = 0$, on $V \times (-T,T) \times (-T,T)$.

q.c.d.

LEMMA 2.2. Let $a(z;t,\tau)$ be as in Lemma 2.1 and let $q(z;t,\tau)$, $a(z;t,\tau)$ be the functions constructed in Lemma 2.1. Then there is a neighborhood $V' \subset V$ of

z_0 such that for every $h(z;t,\tau) \in C^\infty(\mathbb{R}^2 \times M)^{N\times N}$ there are smooth matrix-valued functions $q_1(z;t,\tau)$, $\tilde{c}(z;t,\tau)$ defined on $M \times \mathbb{R}^2$, with $\tilde{c}$ block diagonal $(\tilde{c} = (\tilde{c}_{ii})_{i=1,\ldots,k}$, $\dim \tilde{c}_{ii} = N_i \times N_i)$ so that

$$(2.13) \qquad Lq_1 = (t\,\partial_t - \tau\,\partial_\tau)q_1 - \{aq_1 - q_1\tilde{a}\} = h + q\tilde{c}\ ,$$

on $V' \times (-T,T) \times (-T,T)$ (same T as in Lemma 2.1).

Proof. Once again we assume $k = 2$ and commence to find the formal series for q_1 and $\tilde{c}$, $\sum q_1^{rs}\, t^r\, \tau^s$, $\sum \tilde{c}^{rs}\, t^r\, \tau^r$. Using the Taylor expansion of q and $\tilde{a}$ and taking into account equation (2.13), we see that $q_1^{oo}(z)$ and $\tilde{c}^{oo}(z)$ must satisfy the linear equation :

$$(2.14) \qquad -\{a^{oo}(z)q_1^{oo}(z) - q_1^{oo}(z)\, a^{oo}(z)\} - q^{oo}(z)\,\tilde{c}^{oo}(z) = h^{oo}(z).$$

We let $(\hat{q},\tilde{c}) = \begin{pmatrix} \tilde{c}_{11} & q_{12} \\ q_{21} & \tilde{c}_{22} \end{pmatrix}$ and consider the operator :

$$(2.15) \qquad (\hat{q},\tilde{c}) \longmapsto F_0(z;\hat{q},\tilde{c}) = -\{a^{oo}(z)(\hat{q},0) - (\hat{q},0)\,\tilde{a}^{oo}(z)\} - q^{oo}(z)(0,\tilde{c})\ ,$$

which is surjective at $z = z_0$. In fact, as in Lemma 2.1 we can find a fixed neighborhood $W \subset V$ of z_0 for which the linear operators :

$$(2.16) \qquad (\hat{q},\tilde{c}) \longmapsto F_m(z;\hat{q},\tilde{c}) = m(\hat{q},0) + F_0(z\ ;\hat{q},\tilde{c}),\ m \in \mathbb{Z}\ ,$$

are smoothly invertible on W .

At this neighborhood we solve $F_o(z;\hat{q},\tilde{c}) = h^{oo}(z)$ and let $q_1^{oo}(z) = (\hat{q}(z),0)$, $\tilde{c}^{oo}(z) = (0,\hat{c}(z))$. In the same way we can choose by induction

$$q_1^{rs}(z) = \begin{pmatrix} 0 & q_{12}^{rs}(z) \\ q_{21}^{rs}(z) & 0 \end{pmatrix}, \quad \tilde{c}^{rs}(z) = \begin{pmatrix} \tilde{c}_{11}^{rs}(z) & 0 \\ 0 & \tilde{c}_{22}^{rs}(z) \end{pmatrix},$$

smooth on W such that $(Lq_1 - q\tilde{c})^{rs}(z) = h^{rs}(z)$ on W, for all r,s .

Having constructed the formal power series $\sum q_1^{rs}(z)t^r\tau^s$, $\sum \tilde{c}^{rs}(z)t^r\tau^s$, we proceed exactly as in Lemma 2.1 to conclude the proof.

q.e.d.

The next Lemmas form the core of this Section.

LEMMA 2.3. Let $a(z;t,\tau)$ be a smooth $N \times N$ matrix-valued function defined for $(z;(t,\tau)) \in M \times \mathbb{R}^2$ (M being a C^∞ countable manifold). Assume that for some fixed $\lambda \in \mathbb{C}$ we have :

1) $a(z;t,\tau) = (a_{ij}(z;t,\tau))_{i,j=1,\dots,k}$, where the a_{ij} are $N_i \times N_j$ matrices, $N_1 + \dots + N_k = N, N_i \geq 0$, $\forall i$, $k \geq 2$.

2) At a point $z_o \in M$ we have :

i) $a_{ij}(z_o,0,0) = 0$, for $i \neq j$,

ii) $a_{ii}(z_o,0,0)$, $i = 1,\dots,k$, is in Jordan canonical form with spectrum reduced to th eigenvalue $\lambda - i + 1$.

Then there is a neighborhood $V \times (-T,T) \times (-T,T)$ of $(z_o,0,0)$ $(0 < T \leq 1)$ and two smooth matrix-valued functions $q(z;t,\tau)$, $c(z;t,\tau)$ defined on $M \times \mathbb{R}^2$

such that :

I) $q(z_o;0,0) = I_N$,

II) $$c(z;t,\tau) = \sum_{m=1}^{k-1} e_m(z;\,t\tau)\tau^m + c_o(z;\,t\tau) + \sum_{m=1}^{k-1} d_m(z;\,t\tau)t^m$$

where

α) $c_o(z;s) \in C^\infty(M \times \mathbb{R})^{N\times N}$ is block diagonal with blocks $c_{ii}(z;s)$ of dimension $N_i \times N_i$ and $c_{ii}(z_o;0) = a_{ii}(z_o;0,0)$, $i = 1,\dots,k$.

β) $e_m(z;s) \in C^\infty(M \times \mathbb{R})^{N\times N}$ is of the following form:

$e_m(z;s)$ =

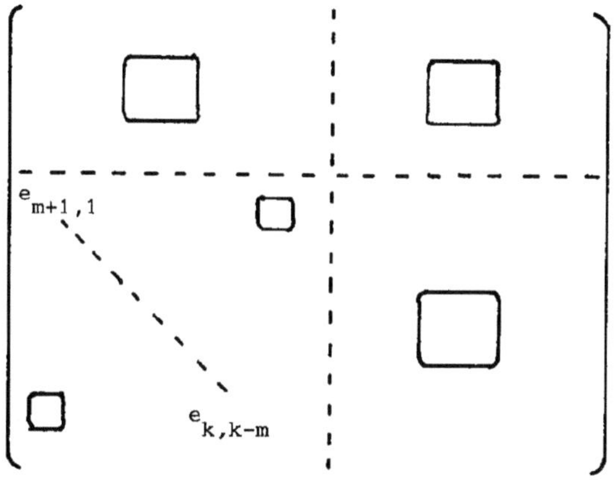

with $e_{m+j,j}(z;s)$ of dimension $N_{m+j} \times N_j$, $j = 1,\dots,k-m$.

γ) $d_m(z;s) \in C^\infty(M \times \mathbb{R})^{N\times N}$ is of the following form:

$d_m(z;s) =$

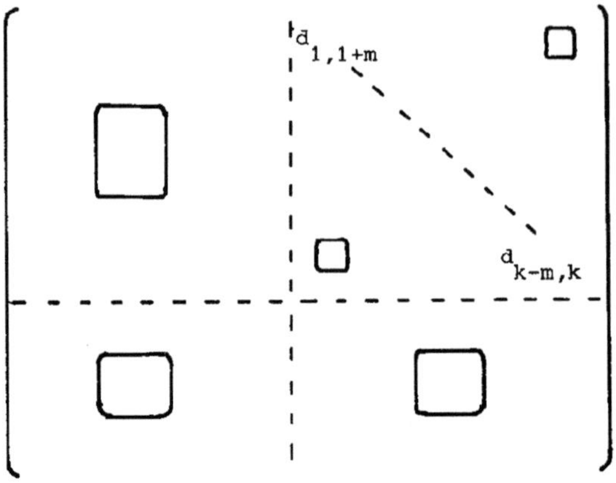

with $d_{i,i+m}(z;s)$ of dimension $N_i \times N_{i+m}$, $i = 1,\dots,k-m$.

III) $Lq = (t\,\partial_t - \tau\partial_\tau)q - \{aq - qc\} = 0,$ on $V \times (-T,T) \times (-T,T)$.

Proof. We assume that the data have Taylor expansion $a(z;t,\tau) \sim \sum a^{rs}(z)t^r\tau^s$ and seek for formal power series $q \sim \sum q^{rs}(z)t^r\tau^s$, $e_m \sim \sum_\ell e_m^\ell(z)(t\tau)^\ell$, $d_m \sim \sum_\ell d_m^\ell(z)(t\tau)^\ell$, $m = 1,\dots,k-1$, and $c_o \sim \sum c_o^\ell(z)(t\tau)^\ell$ such that $(Lq)^{rs}(z) = 0$ for every r,s.

The constant term in the Taylor expansion of Lq is given by

(2.17) $$(Lq)^{oo}(z) = -\{a^{oo}(z)q^{oo}(z) - q^{oo}(z)c_o^o(z)\}.$$

To satisfy $(Lq)^{oo} = 0$ we seek for $q^{oo}(z)$ with $q_{ii}^{oo}(z) = I_{N_i}$, $i = 1,\dots,k$, for every z. It remains to determine $q_{ij}^{oo}(z)$, $i \neq j$, and $c_{ii}^o(z)$.

Let

$$(\hat{q},\hat{c}) = \begin{pmatrix} c_{11} & q_{12} & \cdots & q_{1k} \\ q_{21} & c_{22} & \cdots & q_{2k} \\ \cdots & \cdots & \cdots & \cdots \\ q_{k1} & q_{k2} & \cdots & c_{kk} \end{pmatrix} , \quad q = (\hat{q};I_N) , \quad c = (0,\hat{c}) ,$$

and consider the quadratic equation

$$(2.18) \qquad (\hat{q},\hat{c}) \longmapsto F(z;\hat{q},\hat{c}) = a^{oo}(z)q - qc = 0 .$$

For $z=z_o$, $\hat{q} = 0$ and $\hat{c} = a^{oo}(z_o)$, the derivative $dF_{(\hat{q},\hat{c})}(z_o;\hat{q},\hat{c})$ is given by:

$$(2.19) \qquad \begin{aligned} &(\Delta\hat{q}\, , \Delta\hat{c}) \longmapsto \mathscr{L}(\Delta\hat{q}\, , \Delta\hat{c}) = \\ &= a^{oo}(z_o)(\Delta\hat{q}, 0) - (\Delta\hat{q}\, , 0)a^{oo}(z_o) - (0, \Delta\hat{c}) . \end{aligned}$$

Thus $\mathscr{L} = (\mathscr{L}_{ij})_{i,j=1,\ldots,k}$ with

$$(2.20) \qquad \mathscr{L}_{ij}(\Delta\hat{q}\, , \Delta\hat{c}) = \begin{cases} -\Delta c_{ii} & , \quad i=j , \\ a^{oo}_{ii}(z_o)\,\Delta q_{ij} - \Delta q_{ij}\, a^{oo}_{jj}(z_o), & i \neq j . \end{cases}$$

Since for $i \neq j$, $\sigma(a^{oo}_{ii}(z_o)) \cap \sigma(a^{oo}_{jj}(z_o)) = \emptyset$, we see that $dF_{(\hat{q},\hat{c})}(z_o;\, \hat{q}=0,\, \hat{c}=a^{oo}(z_o))$ is surjective, so that, by the implicit function theorem, we can find a neighborhood W of z_0 and smooth matrix-valued functions $\hat{q}(z)$, $\hat{c}(z)$ defined on W satisfying :

$$(2.21)\qquad \begin{cases} F(z;\hat{q}(z),\ \hat{c}(z)) = 0 & \text{on } W, \\ \hat{q}(z_0) = 0,\quad \hat{c}_{ii}(z_0) = a_{ii}(z_0), & i=1,\dots,k. \end{cases}$$

Then define $q^{oo}(z) = (\hat{q}(z),\ I_N)$ and $c_o^o(z) = (0,\hat{c}(z))$.

We have constructed $q^{oo}(z)$, $c_o^o(z) \in C^\infty(W)$ so that $(Lq)^{oo}(z) = 0$ on W.

To determine the remaining coefficients it is convenient to observe that the coefficient of $t^r\tau^s$ in the Taylor expansion of $q\,c$ is given by :

$$(2.22)\qquad \sum_{m=1}^{\min\{k-1,s\}} \sum_{\ell=0}^{\min\{r,s-m\}} q^{r-\ell,s-m-\ell}(z)e_m^\ell(z) + \sum_{\ell=0}^{\min\{r,s\}} q^{r-\ell,s-\ell}(z)c_o^\ell(z) + \sum_{m=1}^{\min\{k-1,r\}} \sum_{\ell=0}^{\min\{r-m,s\}} q^{r-m-\ell,s-\ell}(z)d_m^\ell(z)\ .$$

For the sake of convenience we define the following linear operators. For $m = 1,\dots,k-1$, put

$$(\hat{q},\hat{d})_m = \begin{pmatrix} q_{11} & q_{12} & \cdots & d_{1,m+1} & \cdots\cdots & q_{1k} \\ q_{21} & q_{22} & \cdots\cdots & d_{2,m+2} & \cdots & q_{2k} \\ \cdots & \cdots & \cdots & \cdots & \cdots & \cdots \\ q_{k-m,1} & q_{k-m,2} & \cdots & \cdots & \cdots & d_{k-m,k} \\ \cdots & \cdots & \cdots & \cdots & \cdots & \cdots \\ q_{k1} & q_{k2} & \cdots & \cdots & \cdots & q_{kk} \end{pmatrix}$$

and define

$$(\hat{q},\hat{d})_m \longmapsto F_m(z;(\hat{q},\hat{d})_m) = m(\hat{q},0)_m - \{a^{oo}(z)(\hat{q},0)_m$$

$$- (\hat{q},0)_m c_o^o(z)\} + q^{oo}(z)(0,\hat{d})_m\ , \quad z \in W.$$

For $m = 1,\dots,k-1$, put

$$(\hat{q},\hat{e})_{-m} = \begin{pmatrix} q_{11} & q_{12} & \cdots & \cdots & q_{1k} \\ \cdots & \cdots & \cdots & \cdots & \cdots \\ e_{m+1,1} & q_{m+1,2} & \cdots & \cdots & q_{m+1,k} \\ q_{m+2,1} & e_{m+2,2} & \cdots & \cdots & q_{m+2,k} \\ \cdots & \cdots & \cdots & \cdots & \cdots \\ q_{k1} & \cdots & e_{k,k-m} & \cdots & q_{kk} \end{pmatrix}$$

and

$$(\hat{q},\hat{e})_{-m} \longmapsto F_{-m}(z;(\hat{q},\hat{e})_{-m}) = -m(\hat{q},0)_{-m} - \{a^{oo}(z)(\hat{q},0)_{-m} - (\hat{q},0)_{-m}\, c_o^o(z)\} + q^{oo}(z)(0,\hat{e})_{-m}\,, \quad z \in W .$$

For $|m| \geq k$ define

$$q = \begin{pmatrix} q_{11} & q_{12} & \cdots & q_{1k} \\ \cdots & \cdots & \cdots & \cdots \\ q_{k1} & q_{k2} & \cdots & q_{kk} \end{pmatrix} \longmapsto F_m(z;q) = mq - \{a^{oo}(z)q - q\, c_o^o(z)\}, \quad z \in W.$$

Finally, for $m = 0$, put

$$(\hat{q},\hat{c})_o = \begin{pmatrix} c_{11} & q_{12} & \cdots & q_{1k} \\ q_{21} & c_{22} & & q_{2k} \\ \cdots & \cdots & \cdots & \cdots \\ q_{k1} & q_{k2} & \cdots & c_{kk} \end{pmatrix}$$

and

$$(\hat{q},\hat{c})_o \longmapsto F_o(z;(\hat{q},\hat{c})_o) = -\{a^{oo}(z)(\hat{q},0)_o - (\hat{q},0)c_o^o(z)\} + q^{oo}(z)(0,\hat{c})_o, \quad z \in W .$$

In the expressions above, $q^{oo}(z)$ and $c_o^o(z)$ are the matrices we have already constructed.

The crucial point to observe is that, as a consequence of hypothesis 2, there is a neighborhood $W' \subset W$ of z_0 such that the operators $F_m(z;\cdot)$ are smoothly invertible for $z \in W'$ and for every $m \in \mathbb{Z}$.

We now determine $q^{ro}(z)$, $d_r^o(z)$, $r = 1,\dots,k-1$. We seek a solution in the form $q^{ro} = (\hat{q},0)_r$, $d_r^o = (0,\hat{d})_r$.

Supposing that $q^{r'o}(z)$ and $d_{r'}^o(z)$, $1 \leq r' < r$, have been already constructed as smooth matrix-valued functions defined on W' so that $(Lq)^{r'o}(z) = 0$, $z \in W'$, using (2.22), we are led to the equation :

$$
\begin{aligned}
(2.23)\qquad & rq^{ro} - \{a^{oo}\, q^{ro} + \sum_{r'=0}^{r-1} a^{r-r',o}\, q^{r'o} - q^{ro}\, c_o^o \\
& - \sum_{r'=0}^{r-1} q^{r-r',o}\, d_{r'}^o\} + q^{oo}\, d_r^o = 0 ,
\end{aligned}
$$

which can be written as:

$$
(2.24)\qquad F_r(z;(\hat{q},\hat{c})_r) = \text{known data} \in C^\infty(W').
$$

Thus $(\hat{q},\hat{d})_r$ are uniquely determined on W' and we define $q^{ro}(z) = (\hat{q}(z),0)_r$, $d_r^o(z) = (0,\hat{d}(z))_r$.

For $r \geq k$ the equation $(Lq)^{ro} = 0$ is of the form $F_r(z;q)$ = known data $\in C^\infty(W')$.

Thus the $q^{ro}(z)$, $r = 1,2,\dots,$ are uniquely determined from the previous construction.

To determine the $q^{os}(z)$, $e_s^o(z)$, $s = 1,\dots,k-1$, we seek a solution in the form $q^{os} = (\hat{q},0)_{-s}$, $e_s^o = (0,\hat{e})_{-s}$. Supposing that $q^{os'}(z)$ and $e_{s'}^o(z)$, $1 \leq s' < s$, have been already constructed as smooth matrices defined on W' so that $(Lq)^{os'}(z) = 0$, $z \in W'$, using (2.22) we are led to the equation:

$$-sq^{os} - \{a^{oo} q^{os} + \sum_{s'=0}^{s-1} a^{o,s-s'} q^{o,s'} - q^{os} c_o^o$$

(2.25)

$$- \sum_{s'=0}^{s-1} q^{o,s-s'} e_{s'}^o\} + q^{oo} e_s^o = 0 ,$$

which can be rewritten as:

(2.26) $$F_{-s}(z;(\hat{q},\hat{e})_{-s}) = \text{known data} \in C^{\infty}(W').$$

Thus, $(\hat{q},\hat{e})_{-s}$ are uniquely determined on W' and we define $q^{os}(z) = (\hat{q}(z),0)_{-s}$, $e_s^o(z) = (0,\hat{e}(z))_{-s}$.

For $s \geqq k$ the equation $(Lq)^{os} = 0$ is of the form $F_{-s}(z,q) = \text{known data} \in C^{\infty}(W')$.

In conclusion we have already determined the q^{ro}, q^{os}, d_r^o, e_s^o for $r,s = 1,2,\ldots$.

To find $q^{11}(z)$ and $c_o^1(z)$ so that $(Lq)^{11}(z) = 0$ on W', we look for $q^{11} = (\hat{q},0)_o$, $c_o^1 = (0,\hat{c})_o$ and we note that we are again led to an equation of the form $F_o(z;(\hat{q},\hat{c})_o) = \text{known data} \in C^{\infty}(W')$.

Knowing $q^{11}(z)$, $c_o^1(z)$ and proceeding as above we determine successively $q^{1r}(z)$, $d_{r-1}^1(z)$, $r = 2,\ldots,k$ and $q^{1r}(z)$ for $r \geqq k+1$, as well as $q^{1s}(z)$, $e_{s-1}^1(z)$, $s = 2,\ldots,k$ and $q^{s1}(z)$ for $s \geqq k+1$.

At this point we can construct $q^{22}(z)$ and $c_o^2(z)$, and so on. Having the formal power series $\sum q^{rs}(z)t^r \tau^s$, $\sum e_m^{\ell}(z) (t\tau)^{\ell}$, $\sum d_m^{\ell}(z) (t\tau)^{\ell}$, $m = 1,\ldots,k-1$, and $\sum c_o^{\ell}(z) (t\tau)^{\ell}$ for $z \in W'$, we proceed exactly as in Lemma 2.1 to conclude the proof.

q.e.d.

LEMMA 2.4. With the notation of Lemma 2.3, there is a neighborhood $V' \subset V$ of z_o such that for every smooth matrix-valued function $h(z;t,\tau)$ defined on $M \times \mathbb{R}^2$ there exist smooth matrices $\tilde{q}(z;t,\tau)$, $\tilde{b}(z;t,\tau)$ defined on $M \times \mathbb{R}^2$

such that :

I) $\tilde{b}$ has the same structure of c in Lemma 2.3, i.e. :

$$\tilde{b}(z;t,\tau) = \sum_{m=1}^{k-1} \tilde{e}_m(z;\ t\tau)\tau^m + \tilde{b}_o(z;\ t\tau) + \sum_{m=1}^{k-1} \tilde{d}_m(z;\ t\tau)t^m.$$

II) $L\tilde{q} = (t\,\partial_t - \tau\,\partial_\tau)\tilde{q} - \{a\tilde{q} - \tilde{q}c\} = h + q\tilde{b}$ on $V' \times (-T,T) \times (-T,T)$,

where q and c are the functions constructed in Lemma 2.3 (with the same T).

Proof. Once again we look for formal power series $\tilde{q} \sim \sum \tilde{q}^{rs}\, t^r\, \tau^s$, $\tilde{e}_m \sim \sum \tilde{e}_m^{\ell}\, s^{\ell}$, $\tilde{b}_o \sim \sum \tilde{b}_o^{\ell}\, s^{\ell}$, $\tilde{d}_m \sim \sum \tilde{d}_m^{\ell}\, s^{\ell}$, $m = 1,\ldots,k-1$, such that $(L\tilde{q} - q\tilde{b})^{rs} = h^{rs}$ for every r,s. When $r = s = 0$, $\tilde{q}^{oo}(z)$ and $\tilde{b}_o^o(z)$ must satisfy the linear equation

$$(2.27) \qquad -\{a^{oo}(z)\tilde{q}^{oo}(z) - \tilde{q}^{oo}(z)c_o^o(z)\} - q^{oo}(z)\tilde{b}_o^o(z) = h^{oo}(z).$$

A solution may be found in the form $\tilde{q}^{oo} = (\hat{q},0)_o$, $\tilde{b}_o^o = (0,\hat{b})_o$ (with the notation introduced in the proof of Lemma 2.3).

Using the same kind of induction as in the proof of Lemma 2.3, algebraic equations of the form

$$(2.28) \qquad F_m(z;(\hat{q},\hat{b})_m) = \text{known data}, \quad m \in \mathbb{Z},$$

can be smoothly solved in a fixed neighborhood $V' \subset V$ of z_o .

After the formal series have been constructed, we conclude as in Lemma 2.1.

q.e.d.

Remarks. 1) If the original matrix a (in Lemmas 2.3, 2.4) is block lower triangular we can choose q and c in Lemma 2.3 to be block lower triangular (so that $d_m = 0$, $m = 1,\dots,k-1$. Moreover, if in addition the data h in Lemma 2.4 is block lower triangular we can choose $\tilde{q}$ and $\tilde{b}$ in Lemma 2.4 to be block lower triangular (so that $\tilde{d}_m = 0$, $m = 1,\dots,k-1$).

2) We explicitly note that, as it follows from the above proofs, when $k = 1$, Lemmas 2.3 and 2.4 hold with $c = c_o$ and $b = b_o$ respectively.

Lemmas 2.1 - 2.4 provide the necessary tools in order to construct the symbols of intertwining p d o's which reduce the system (2.1) either to a system of the form studied in Section 1 or to some enlarged system.

Consider the (matrix) operator $A(t,x,D_t,D_x) \in L^o(\tilde{X};N \times N)$ and denote by $a_o(t,x;\tau,\xi)$ its principal symbol.

Without loss of generality we can suppose that the matrix $a_o(\rho_o)$, $\rho_o = (0,x_o,0,\xi^{(o)})$ $\in \Sigma_o$, has the following structure:

i) $a_o(\rho_o) = (a_{ii})_{i=1,\dots,k}$, $k \geq 1$, is block diagonal with blocks of dimension $N_i \times N_i$, $N_1+\dots+N_k = N$.

ii) $\sigma(a_{ii} + r\, I_{N_i}) \cap \sigma(a_{jj}) = \emptyset \quad \begin{cases} i,j = 1,\dots,k,\ i \neq j \\ r \in \mathbb{Z} \end{cases}$

(this condition is empty if $k = 1$).

We have the following result :

PROPOSITION 2.1. Suppose that $a_o(\rho_o)$ has the above structure with $k \geqq 2$. Then there exist :

1) A conic neighborhood V_{ρ_o} of ρ_o, $V_{\rho_o} \cap N^*X = \emptyset$;

2) A p d o $Q \in L^o(\tilde{X};\ N\times N)$ elliptic in V_{ρ_o} ;

3) An operator $\tilde{A} \in L^o(\tilde{X};\ N\times N)$ which is block diagonal, i.e.

$\tilde{A} = (\tilde{A}_{ii})_{i=1,\dots,k}$ with $\tilde{A}_{ii} \in L^o(\tilde{X};\ N_i \times N_i)$, $i = 1,\dots,k$;

such that

i) The principal symbol $\tilde{a}_{ii}^{(o)}$ of $\tilde{A}_{ii}$ satisfies $\tilde{a}_{ii}^{(o)}(\rho_o) = a_{ii}$, $i = 1,\dots,k$.

ii) $(t\,\partial_t I_N - A)Q \equiv Q(t\,\partial_t I_N - \tilde{A})$ in V_{ρ_o}.

Proof. Suppose $|\xi^{(o)}| = 1$ and for every $\varepsilon > 0$ define the sets

$$\Lambda^{\varepsilon}_{\rho_o} = \{(t,x;\tau,\xi') \in \tilde{X}\times \mathbb{R}_{\tau} \times S^{n-1} \mid |\xi' - \xi^{(o)}| < \varepsilon,\ |x - x_o| < \varepsilon,\ |t|,\ |\tau| < \varepsilon\},$$

$$V^{\varepsilon}_{\rho_o} = \{(t,x;\tau,\xi) \in \dot{T}^*X \setminus N^*X \mid (t,x;\ \frac{\tau}{|\xi|},\ \frac{\xi}{|\xi|}) \in \Lambda^{\varepsilon}_{\rho_o}\}\ .$$

Letting $Q,\tilde{A} \in L^o(\tilde{X};N\times N)$ with symbols $\sum_{j\geq 0} q_{-j}$, $\sum_{j\geq 0} \tilde{a}_{-j}$ respectively, and imposing the condition $(t\,\partial_t I_N - A)Q \equiv Q(t\,\partial_t I_N - \tilde{A})$, we obtain a sequence of transport equations involving the q_{-j} and $\tilde{a}_{-j}$, $j \geq 0$. Precisely, at the principal symbol level, we have the system:

$$(2.28) \qquad (t\,\partial_t - \tau\,\partial_\tau)q_o - \{a_o\, q_o - q_o\, \tilde{a}_o\} = 0\ .$$

Using Lemma 2.1, with $M = X \times S^{n-1}$, $z = (x,\xi')$, we know that there exist two smooth matrix-valued functions $\hat{q}_0(t,x,\tau,\xi')$, $\hat{a}_0(t,x,\tau,\xi')$ defined on $\mathbb{R}^2_{(t,\tau)} \times M$, with $\hat{a}_0$ block diagonal, such that:

$$(2.29)\qquad \begin{aligned} &(t\,\partial_t - \tau\partial_\tau)\hat{q}_0(t,x;\tau,\xi') = \\ &= a_0(t,x;\tau,\xi')\hat{q}_0(t,x;\tau,\xi') - \hat{q}_0(t,x;\tau,\xi')a_0(t,x;\tau,\xi') \end{aligned}$$

on some $\Lambda^{\varepsilon}_{\rho_0}$. Moreover, we have $\hat{q}_0(\rho_0) = I_N$ and $\hat{a}_0(\rho_0) = a_0(\rho_0)$.
On $\dot{T}^*\tilde{X} \times N^*X$ define :

$$(2.30)\qquad \begin{cases} q_0(t,x,\tau,\xi) = \hat{q}_0(t,x, \dfrac{\tau}{|\xi|}, \dfrac{\xi}{|\xi|}), \\ \tilde{a}_0(t,x,\tau,\xi) = \hat{a}_0(t,x, \dfrac{\tau}{|\xi|}, \dfrac{\xi}{|\xi|}) . \end{cases}$$

By suitably modifying q_0, $\tilde{a}_0$ out of $V^{2\varepsilon}_{\rho_0}$ we can suppose that $q_0,\tilde{a}_0$ are defined on $\dot{T}^*\tilde{X}$ as smooth matrix-valued functions positively homogeneous of degree 0 in (τ,ξ) and with $\tilde{a}_0$ block diagonal. From (2.30), (2.29), it follows that (2.28) is satisfied in $V^{\varepsilon}_{\rho_0}$. Therefore we choose Q and A with principal symbols $q_0,\tilde{a}_0$ as constructed above. For the terms q_{-1}, a_{-1} we have another system of the form:

$$(2.31)\qquad (t\,\partial_t - \tau\partial_\tau)q_{-1} - \{a_0\, q_{-1} - q_{-1}\, \tilde{a}_0\} = q\, \tilde{a}_{-1} + g_{-1} ,$$

where $g_{-1} = a_{-1}\, q_0 - q_{-1}\, \tilde{a}_0 + \frac{1}{i}\{\partial_{t,x}\, a_0\quad \partial_{\tau,\xi}\, q_0 - \partial_{t,x}\, q_0\quad \partial_{\tau,\xi}\, \tilde{a}_0\}$.

Using Lemma 2.2, there exist two smooth matrix-valued functions $\hat{q}_{-1}(t,x,\tau,\xi')$, $\hat{a}_{-1}(t,x,\tau,\xi')$ defined on $\mathbb{R}^2_{(t,\tau)} \times M$, with $\hat{a}_{-1}$ block diagonal, such that:

$$(t\,\partial_t - \tau\partial_\tau)\hat{q}_{-1}(t,x,\tau,\xi') - \{a_0(t,x,\tau,\xi')\hat{q}_{-1}(t,x,\tau,\xi')$$

$$(2.32)\qquad - \hat{q}_{-1}(t,x,\tau,\xi')\hat{a}_0(t,x,\tau,\xi')\} =$$

$$= \hat{q}_0(t,x,\tau,\xi')\hat{a}_{-1}(t,x,\tau,\xi') + g_{-1}(t,x,\tau,\xi')$$

on some $\Lambda^{\varepsilon'}_{\rho_0}$, $\varepsilon' \leqq \varepsilon$. Define on $\dot{T}^*\tilde{X} \setminus N^*X$:

$$(2.33)\qquad \begin{cases} q_{-1}(t,x,\tau,\xi) = |\xi|^{-1}\,\hat{q}_{-1}(t,x,\dfrac{\tau}{|\xi|},\dfrac{\xi}{|\xi|})\,, \\ \tilde{a}_{-1}(t,x,\tau,\xi) = |\xi|^{-1}\,\hat{a}_{-1}(t,x,\dfrac{\tau}{|\xi|},\dfrac{\xi}{|\xi|})\,. \end{cases}$$

Modifying q_{-1} , $\tilde{a}_{-1}$ out of $V^{2\varepsilon'}_{\rho_o}$ we can suppose that q_{-1}, $\tilde{a}_{-1}$ are smoothly defined on $\dot{T}^*\tilde{X}$ and positively homogeneous of degree -1 in (τ,ξ). From (2.33), (2.32) it follows that (2.31) is satisfied in $V^{\varepsilon'}_{\rho_0}$. Therefore we choose Q and A such that $Q - q_0(t,x,D_t,D_x) - q_{-1}(t,x,D_t,D_x)$ and $\tilde{A} - \tilde{a}_0(t,x,D_t,D_x) - \tilde{a}_{-1}(t,x,D_t,D_x)$ are in $L^{-2}(\tilde{X};\, N\times N)$. Going on we obtain a transport equation for q_{-2}, $\tilde{a}_{-2}$ which can be solved using Lemma 2.2 in the same set $\Lambda^{\varepsilon'}_{\rho_0}$, and so on. Defining $Q \sim \sum_{j\geq 0} q_{-j}(t,x,D_t,D_x)$, $\tilde{A} \sim \sum_{j\geq 0} \tilde{a}_{-j}(t,x,D_t,D_x)$ we are done.

q.e.d.

The preceding result reduces the study of system (2.1) near Σ_0 to the study of a finite number of decoupled systems $t\,\partial_t I_{N_i} - \tilde{A}_{ii}(t,x,D_t,D_x)$, $i = 1,\ldots,k$, such that denoting by $\tilde{a}^{(o)}_{ii}(t,x,\tau,\xi)$ the principal symbol of $\tilde{A}_{ii}$, we have at $\rho_0 \in \Sigma_0$:

1) $\sigma(\tilde{a}_{ii}^{(o)}(\rho_o) + r\, I_{N_i}) \cap \sigma(\tilde{a}_{jj}^{(o)}(\rho_o)) = \emptyset$,

for $i,j = 1,\dots,k$, $i \neq j$, and for every $r \in \mathbb{Z}$.

2) $\tilde{a}_{ii}^{(o)}(\rho_o)$ is in Jordan canonical form. Precisely, either :

$$(2.34)\quad \tilde{a}_{ii}^{(o)}(\rho_o) = \left(\begin{array}{c|c|c|c} \begin{matrix} \lambda_i & & * \\ & \ddots & \\ 0 & & \lambda_i \end{matrix} & & & 0 \\ \hline & \begin{matrix} \lambda_i - \ell_{1i} & & * \\ & \ddots & \\ 0 & & \lambda_i - \ell_{1i} \end{matrix} & & \\ \hline & & \ddots & \\ \hline 0 & & & \begin{matrix} \lambda_i - \ell_{h_i i} & & * \\ & \ddots & \\ 0 & & \lambda_i - \ell_{h_i i} \end{matrix} \end{array}\right) , \quad i=1,\dots,k,$$

where $\lambda_i \in \mathbb{C}$ and $\ell_{ji} \in N$, $j = 1,\dots,h_i$ with $1 \leq \ell_{1i} < \ell_{2i} < \dots < \ell_{h_i i}$, $h_i \geq 1$; or :

$$(2.35)\quad \tilde{a}_{ii}^{(o)}(\rho_o) = \begin{pmatrix} \lambda_i & & & * \\ & \lambda_i & & \\ & & \ddots & \\ 0 & & & \lambda_i \end{pmatrix} .$$

where $\lambda_i \in \mathbb{C}$.

We shall now study the systems $t\partial_t I_{N_i} - \tilde{A}_{ii}(t,x,D_t,D_x)$ more closely. To simplify the notation we drop the indices i and consider a system:

(2.36) $\qquad (t\partial_t I_N - A(t,x,D_t,D_x))u = f$,

where at some point $\rho_o \in \Sigma_o$ the principal symbol $a_o(t,x,\tau,\xi)$ of the operator $A \in L^o(\tilde{X}; N \times N)$ is in Jordan canonical form. Precisely, either :

$$(2.37) \quad a_o(\rho_o) = \begin{pmatrix} \begin{array}{cc} \lambda & * \\ & \ddots \\ & & \lambda \end{array} & & & \square \\ & \begin{array}{cc} \lambda-\ell_1 & * \\ & \ddots \\ & & \lambda-\ell_1 \end{array} & & \\ & & \ddots & \\ \square & & & \begin{array}{cc} \lambda-\ell_h & * \\ & \ddots \\ & & \lambda-\ell_h \end{array} \end{pmatrix},$$

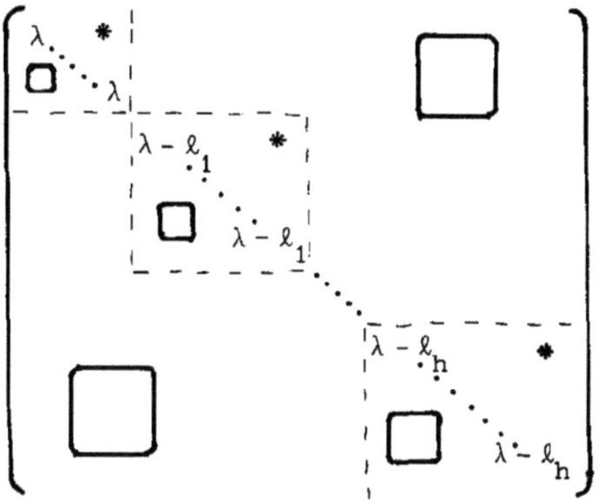

where $\lambda \in \mathbb{C}$, $\ell_1,\ldots,\ell_h \in \mathbb{N}$ with $1 \leq \ell_1 < \ldots < \ell_h$, $h \geq 1$; or:

$$(2.38) \quad a_o(\rho_o) = \begin{pmatrix} \lambda & & & * \\ & \lambda & & \\ & & \ddots & \\ \square & & & \lambda \end{pmatrix},$$

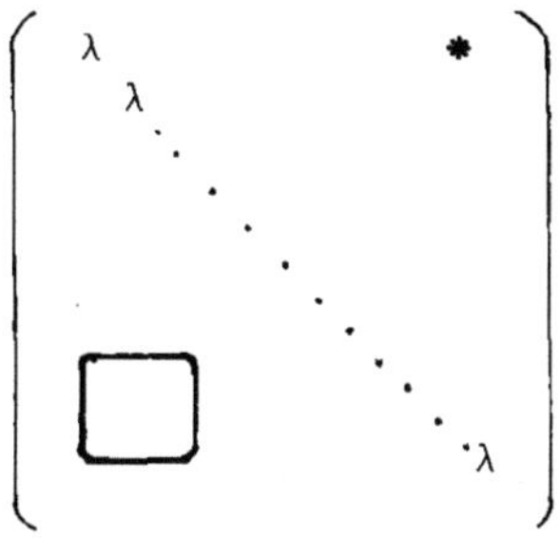

where $\lambda \in \mathbb{C}$.

First we make the following simplifying remark. If $a_o(\rho_o)$ has the form (2.37), letting $\ell_h = k-1$ we can suppose without loss of generality that $a_o(\rho_o)$ is block diagonal : $a_o(\rho_o) = (a_{ii}^{(o)}(\rho_o))_{i=1,\dots,k}$, where $a_{ii}^{(o)}(\rho_o)$ has dimension $N_i \times N_i$, $N_1 + \dots + N_k = N$, and $a_{ii}^{(o)}(\rho_o)$ is in Jordan canonical form with spectrum reduced to the eigenvalue $\lambda - i + 1$, $i = 1,\dots,k$. To obtain this structure it is enough to add to the original system (2.36) new scalar equations of the form

$$(2.39) \qquad t\partial_t u_j - (\lambda - j) \quad u_j = f_j, \quad u_j, f_j \in C^\infty(\tilde{X}),$$

corresponding to the $j \in \{1,2,\dots,k-1\} \setminus \{\ell_1, \ell_2, \dots, \ell_h\}$.

Therefore, we shall suppose that either $a_o(\rho_o)$ has the form (2.38) or the form (2.37) with $\ell_1 = 1$, $\ell_2 = 2,\dots,\ell_h = k-1$, and each block

$$\begin{pmatrix} \lambda - i+1 & & * \\ & \ddots & \\ \square & & \lambda - i+1 \end{pmatrix} \quad \text{has dimension} \quad N_i \times N_i, \; i = 1,\dots,k.$$

We can now state the crucial result of this Section.

PROPOSITION 2.2. Given the system (2.36), suppose that $a_o(\rho_o)$ has the form (2.37) with $\ell_1 = 1$, $\ell_2 = 2,\dots,\ell_{k-1} = k-1$, for some $k \geq 2$, and

$$\begin{pmatrix} \lambda - i+1 & & * \\ & \ddots & \\ \square & & \lambda - i+1 \end{pmatrix} \quad \text{have dimension} \quad N_i \times N_i, \; i = 1,\dots,k.$$

Then there exist :

1) A conic neighborhood V_{ρ_o} of ρ_o, $V_{\rho_o} \cap N^*X = \emptyset$;

2) Two p d o's $Q_1, Q_2 \in L^o(\tilde{X}; N \times N)$ elliptic in V_{ρ_o} ;

3) A pdo $C \in L^{o}(\tilde{X};\ N \times N)$ with the following matrix structure :

$$C(t,x,D_t,D_x) = \sum_{m=1}^{k-1} E_m(x,D_x)D_t^m + C_o(x,D_x) +$$

$$+ \sum_{m=1}^{k-1} D_m(x,D_x)t^m \ , \quad \text{where}$$

α) $C_o(x,D_x) \in L^{o}(X;\ N \times N)$ is block diagonal with blocks $C_{ii}(x,D_x) \in$ $\in L^{o}(X;\ N_i \times N_i)$, such that the principal symbol $c_o^{(o)}(x,\xi)$ of C_o satisfies:

$$c_o^{(o)}(x_o,\xi^{(o)}) = a_o(\rho_o).$$

β) $E_m(x,D_x) \in L^{-m}(X;\ N \times N)$ is of the following form :

$E_m(x,D_x) =$

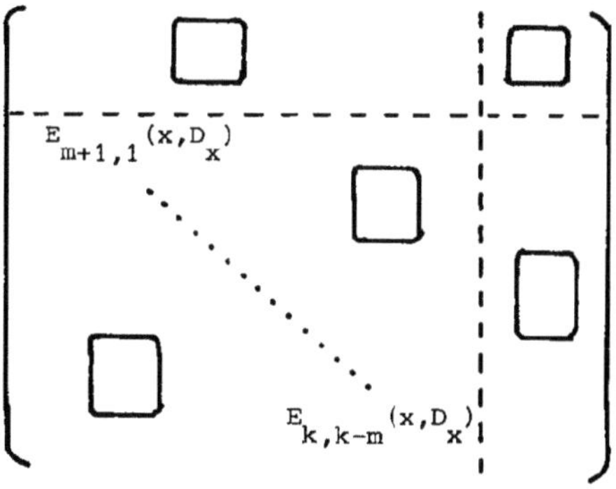
E_{m+1,1}(x,D_x)
E_{k,k-m}(x,D_x)

with $E_{m+j,j}(x,D_x) \in L^{-m}(X;\ N_{m+j} \times N_j)$, $j = 1,\dots,k-m$.

γ) $D_m(x,D_x) \in L^{o}(X;\ N \times N)$ is of the following form :

$$D_m(x,D_x) = \left(\begin{array}{c|c} \square & \begin{array}{ccc} D_{1,1+m}(x,D_x) & & \square \\ & \ddots & \\ \square & & D_{k-m,k}(x,D_x) \end{array} \\ \hline \square & \square \end{array}\right)$$

with $D_{i,i+m}(x,D_x) \in L^{o}(X;\ N_i \times N_{i+m})$, $i = 1,\dots,k-m$, such that

(2.40) $$(t\,\partial_t I_N - A)Q_1 \equiv Q_2(t\,\partial_t I_N - C) \quad \text{in } V_{\rho_o}\,.$$

If $a_o(\rho_o)$ has the form (2.38), we can take $C = C_o(x,D_x)$ and (2.40) is satisfied.

<u>Proof.</u> Suppose we have already found a pdo $Q \in L^{o}(\tilde{X};\ N \times N)$, elliptic in a conic neighborhood of ρ_o, such that on this neighborhood we have $(t\,\partial_t I_N - A)Q \equiv$ $\equiv Q(t\,\partial_t I_N - \tilde{C})$ with an operator $\tilde{C} \in L^{o}(\tilde{X};\ N \times N)$ for which $\tilde{C}(t,x,D_t,D_x) =$

$$= \sum_{m=1}^{k-1} \tilde{E}_m(t,x,D_t,D_x)D_t^m + \tilde{C}_o(t,x,D_t,D_x) + \sum_{m=1}^{k-1} \tilde{D}_m(t,x,D_t,D_x)t^m\,,$$ where the operators $\tilde{E}_m$, $\tilde{C}_o$, $\tilde{D}_m$ have the same matrix structure of the E_m, C_o, D_m, respectively, but with full symbols depending only on x,ξ and $t\tau$ in a conic neighborhood of ρ_o, and $\tilde{c}_o^{(o)}(x_o,\xi^{(o)},0) = a_o(\rho_o)$.

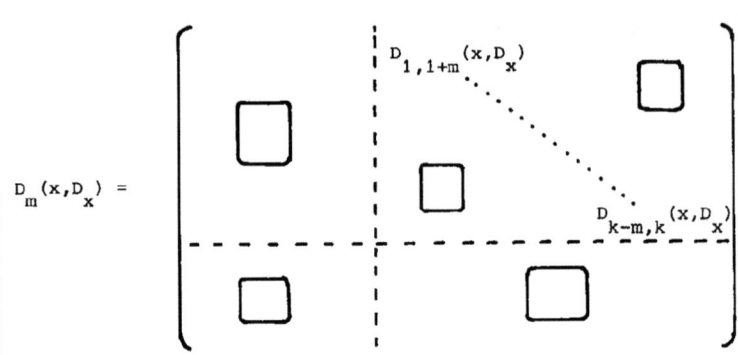

We claim that there exists a pdo $F \in L^{o}(\tilde{X};\ N \times N)$ elliptic near ρ_o, with full symbol depending only on x,ξ, $t\tau$, such that $(t\partial_t I_N - \tilde{C})F \equiv (t\partial_t I_N - C)$ (near ρ_o) with C as in the statement of Proposition 2.2.

For this use a trick of Chazarain [8]. Writing $F \sim I_N + F_{-1} + \dots$, and $C \sim C^{(o)} + C^{(-1)} + \dots$, we first want $(t\partial_t I_N - \tilde{C})(I_N + F_{-1}) = t\partial_t I_N - C^{(o)}$ mod. $L^{-1}(\tilde{X}; N \times N)$.

On the principal symbol level we have in a neighborhood of ρ_o where $\xi \neq 0$:

$$
\begin{aligned}
& it\tau\, f_{-1}(t,x,\tau,\xi) - \Big\{ \sum_{m=1}^{k-1} \tilde{e}_m^{(o)}(x,\xi,t\tau)\tau^m + \tilde{c}_o^{(o)}(x,\xi,t\tau) \\
& + \sum_{m=1}^{k-1} \tilde{d}_m^{(o)}(x,\xi,t\tau)t^m \Big\} = - \Big\{ \sum_{m=1}^{k-1} e_m^{(o)}(x,\xi)\tau^m + c_o^{(o)}(x,\xi) \\
& + \sum_{m=1}^{k-1} d_m^{(o)}(x,\xi)t^m \Big\} .
\end{aligned} \tag{2.41}
$$

Define

$$
\left\{
\begin{aligned}
& e_m^{(o)}(x,\xi) = \tilde{e}_m^{(o)}(x,\xi,0),\quad d_m^{(o)}(x,\xi) = \tilde{d}_m^{(o)}(x,\xi,0),\quad m = 1,\dots,k-1. \\
& c_o^{(o)}(x,\xi) = \tilde{c}_o^{(o)}(x,\xi,0) \\
& f_{-1}(t,x,\tau,\xi) = \frac{1}{it\tau}\Big\{ \sum_{m=1}^{k-1} (\tilde{e}_m^{(o)}(x,\xi,t\tau) - \tilde{e}_m^{(o)}(x,\xi,0))\tau^m \\
& + \tilde{c}_o^{(o)}(x,\xi,t\tau) - \tilde{c}_o^{(o)}(x,\xi,0) + \sum_{m=1}^{k-1} (\tilde{d}_m^{(o)}(x,\xi,t\tau) - \tilde{d}_m^{(o)}(x,\xi,0))t^m \Big\}
\end{aligned}
\right. \tag{2.42}
$$

Observe that f_{-1} is smooth at $t\tau = 0$ and is positively homogeneous of degree -1 in (τ,ξ).

Assume by induction that

$$(t\partial_t I_N - \tilde{C})(I_N + F_{-1} + \ldots + F_{-\ell}) \equiv t\partial_t I_N - C^{(0)} - C^{(-1)} - \ldots - C^{(-\ell+1)} ,$$

mod $L^{-\ell}(\tilde{X};\ N \times N)$ (obviously we require that the $C^{(-j)} \in L^{-j}(\tilde{X};\ N \times N)$ have the same matrix structure of C). We want to find $F_{-\ell-1} \in L^{-\ell-1}$ and $C^{(-\ell)} \in L^{-\ell}$ so that :

$$(t\partial_t I_N - \tilde{C})\ (I_N + F_{-1} + \ldots + F_{-\ell-1}) \equiv t\partial_t I_N - C^{(0)} - \ldots - C^{(-\ell)}$$

mod $L^{-\ell-1}(\tilde{X};\ N \times N)$. Hence $(t\partial_t I_N - \tilde{C})F_{-\ell-1} = C^{(-\ell)} + G_{-\ell}$ mod $L^{-\ell-1}(\tilde{X};\ N \times N)$, where $G_{-\ell} \in L^{-\ell}(\tilde{X};\ N \times N)$ by induction, and the principal symbol of $G_{-\ell}$ has the following structure :

$$g_{-\ell}(t,x,\tau,\xi) = \sum_{m=1}^{k-1} \varphi_m(x,\xi,t\tau)\tau^m + \chi(x,\xi,t\tau) + \tag{2.43}$$
$$+ \sum_{m=1}^{k-1} \psi_m(x,\xi,t\tau)t^m,$$

where φ_m, ψ_m and χ have the same matrix structure of E_m, D_m and C_o in C, the φ_m are positively homogeneous of degree $-m-\ell$, while χ and the ψ_m are positively homogeneous of degree $-\ell$.

To have $(t\partial_t I_N - \tilde{C})F_{-\ell-1} \equiv C^{(-\ell)} + G_{-\ell}$ mod $L^{-\ell-1}(\tilde{X};\ N \times N)$ we must equate the principal symbol of both sides and obtain :

$$(2.44)\quad \begin{aligned} it\tau\, f_{-\ell-1}(t,x,\tau,\xi) &= \sum_{m=1}^{k-1} [\varphi_m(x,\xi,t\tau) + e_m^{(-\ell)}(x,\xi)]\,\tau^m + \\ &+ c_o^{(-\ell)}(x,\xi) + \chi(x,\xi,t\tau) + \sum_{m=1}^{k-1} [\psi_m(x,\xi,t\tau) + d_m^{(-\ell)}(x,\xi)]\,t^m . \end{aligned}$$

Define

$$(2.45)\quad \begin{cases} e_m^{(-\ell)}(x,\xi) = -\varphi_m(x,\xi,0)\,, \quad d_m^{(-\ell)}(x,\xi) = -\psi_m(x,\xi,0)\,, \quad m = 1,\dots,k-1, \\ c_o^{(-\ell)}(x,\xi) = -\chi(x,\xi,0) \end{cases}$$

and then define $f_{-\ell-1}(t,x,\tau,\xi)$ by dividing the r.h.s. of (2.44) by $it\tau$. The induction is completed. Defining $Q_1 = QF$, $Q_2 = Q$ we are done. Thus it remains to prove that we can find an operator $Q \in L^o(\tilde{X};\ N\times N)$ elliptic near ρ_o such that in a conic neighborhood of ρ_o we have $(t\,\partial_t I_N - A)Q \equiv Q(t\,\partial_t I_N - \tilde{C})$ with an operator $\tilde{C} = \sum_{m=1}^{k-1} \tilde{E}_m D_t^m + \tilde{C}_o + \sum_{m=1}^{k-1} \tilde{D}_m t^m$ having the same matrix structure of C but with the full symbols of $\tilde{E}_m, \tilde{D}_m$ and $\tilde{C}_o$ depending only on $x,\xi,t\tau$ (near ρ_o), and such that $\tilde{c}_o^{(o)}(x_o,\ \xi^{(o)},0) = a_o(\rho_o)$.

Letting $Q \in L^o(\tilde{X};\ N\times N)$ and $\tilde{C} \in L^o(X;\ N\times N)$, $\tilde{C} = \sum_{m=1}^{k-1} \tilde{E}_m D_t^m + \tilde{C}_o + \sum_{m=1}^{k-1} \tilde{D}_m t^m$, with symbols $q \sim \sum_{j\geq o} q_{-j}$, $\tilde{e}_m(x,\xi,t\tau) \sim \sum_{j\geq o} \tilde{e}_m^{(-j)}(x,\xi,t\tau)$, $\tilde{c}_o(x,\xi,t\tau) \sim \sum_{j\geq o} \tilde{c}_o^{(-j)}(x,\xi,t\tau)$, $\tilde{d}_m(x,\xi,t\tau) \sim \sum_{j\geq o} \tilde{d}_m^{(-j)}(x,\xi,t\tau)$, we impose that $(t\,\partial_t I_N - A)Q \equiv Q(t\,\partial_t I_N - \tilde{C})$ in a conic neighborhood of ρ_o. We thus obtain a sequence of transport equations involving the above symbols . Precisely, at the principal symbol level, we have the system :

$$(2.46) \qquad (t\,\partial_t - \tau\,\partial_\tau) q_o - \{a_o\, q_o - q_o\, \tilde{c}^{(o)}\} = 0 ,$$

where $\tilde{c}^{(o)} = \sum_{m=1}^{k-1} \tilde{e}_m^{(o)} \tau^m + \tilde{c}_o^{(o)} + \sum_{m=1}^{k-1} \tilde{d}_m^{(o)} t^m$ denotes the principal symbol of $\tilde{C}$.

Using Lemma 2.3 with $M = X \times S^{n-1}$, $z = (x,\xi')$, we know that there exist two smooth matrix valued functions $\hat{q}_o(t,x,\tau,\xi')$, $\hat{c}^{(o)}(t,x,\tau,\xi') =$

$= \sum_{m=1}^{k-1} \hat{e}_m^{(o)}(x,\xi';\, t\tau)\tau^m + \hat{c}_o^{(o)}(x,\xi';\, t\tau) + \sum_{m=1}^{k-1} \hat{d}_m^{(o)}(x,\xi';\, t\tau) t^m$, having the correct matrix structure, defined on $M \times \mathbb{R}^2$, such that

$$(2.47) \qquad \begin{aligned} &(t\,\partial_t - \tau\,\partial_\tau)\hat{q}_o(t,x,\tau,\xi') = \\ &= \{a_o(t,x,\tau,\xi')\hat{q}_o(t,x,\tau,\xi') - \hat{q}_o(t,x,\tau,\xi')\hat{c}^{(o)}(t,x,\tau,\xi')\} , \end{aligned}$$

on some $\Lambda^{\varepsilon}_{\rho_o}$ (we use the notation $\Lambda^{\varepsilon}_{\rho_o}$, $V^{\varepsilon}_{\rho_o}$ as in Proposition 2.1). Moreover, $\hat{q}_o(0,x_o;0,\xi^{(o)}) = I_N$ and $\hat{c}_o^{(o)}(x_o,\xi^{(o)};\, 0) = a_o(\rho_o)$.

Define:

$$(2.48) \qquad \begin{cases} q_o(t,x,\tau,\xi) = \hat{q}_o\left(t,x,\dfrac{\tau}{|\xi|},\dfrac{\xi}{|\xi|}\right) , \\ \tilde{c}^{(o)}(t,x,\tau,\xi) = \displaystyle\sum_{m=1}^{k-1} \left(\left(\hat{e}_m^{(o)}\left(x,\dfrac{\xi}{|\xi|},\dfrac{t\tau}{|\xi|}\right)|\xi|^{-m}\right)\tau^m\right. + \\ + \hat{c}_o^{(o)}\left(x,\dfrac{\xi}{|\xi|},\dfrac{t\tau}{|\xi|}\right) + \displaystyle\sum_{m=1}^{k-1} \hat{d}_m^{(o)}\left(x,\dfrac{\xi}{|\xi|},\dfrac{t\tau}{|\xi|}\right) t^m . \end{cases}$$

By suitably modifying q_o and $\tilde{c}^{(o)}$ out of $V^{2\varepsilon}_{\rho_o}$ we can suppose that q_o, $\tilde{c}^{(o)}$ are smoothly defined on $\dot{T}^*\tilde{X}$ (preserving the matrix structure). From (2.48), (2.47) it follows that equation (2.46) is satisfied in $V^{\varepsilon}_{\rho_o}$. Therefore we choose Q with principal symbol q_o and $\tilde{C}$ with principal symbol $\sum_{m=1}^{k-1} \tilde{e}^{(o)}_m (x,\xi,t\tau)\tau^m +$ $\tilde{c}^{(o)}_o (x,\xi,t\tau) + \sum_{m=1}^{k-1} \tilde{d}^{(o)}_m (x,\xi,t\tau)t^m$, where $\tilde{e}^{(o)}_m (x,\xi,t\tau) = \hat{e}^{(o)}_m (x, \frac{\xi}{|\xi|}, t\frac{\tau}{|\xi|})|\xi|^{-m}$ $\tilde{d}^{(o)}_m (x,\xi,t\tau) = \hat{d}^{(o)}_m (x, \frac{\xi}{|\xi|}, t\frac{\tau}{|\xi|})$, $m = 1,\dots,k-1$, $\tilde{c}^{(o)}_o (x,\xi,t\tau) =$

$= \hat{c}^{(o)}_o (x, \frac{\xi}{|\xi|}, t\frac{\tau}{|\xi|})$ on $V^{2\varepsilon}_{\rho_o}$.

For the terms q_{-1}, $c^{(-1)}$ we have another system of the form :

$$(t\,\partial_t - \tau\,\partial_\tau)q_{-1} - \{a_o\, q_{-1} - q_{-1}\,\tilde{c}^{(o)}\} = q_o\,\tilde{c}^{(-1)} + g_{-1}\,, \tag{2.49}$$

where $g_{-1} = a_{-1}\, q_o - q_{-1}\,\tilde{c}^{(o)} + \frac{1}{i}\{\partial_{t,x}\, a_o \cdot \partial_{\tau,\xi}\, q_o - \partial_{t,x}\, q_o \cdot \partial_{\tau,\xi}\, \tilde{c}^{(o)}\}$.

Using Lemma 2.4 we know that there exist two smooth matrix-valued functions

$\hat{q}_{-1}(t,x,\tau,\xi')$, $\hat{c}^{(-1)}(t,x,\tau,\xi') = \sum_{m=1}^{k-1} \hat{e}^{(-1)}_m (x,\tau';\ t\tau)\tau^m + \hat{c}^{(-1)}_o (x,\xi';\ t\tau) +$ $+ \sum_{m=1}^{k-1} \hat{d}^{(-1)}_m (x,\xi';\ t\tau)t^m$, having the correct matrix structure, defined on $M \times \mathbb{R}^2$, and such that

$$\begin{aligned}&(t\,\partial_t - \tau\,\partial_\tau)\hat{q}_{-1}(t,x,\tau,\xi') - \{a_o(t,x,\tau,\xi')\hat{q}_{-1}(t,x,\tau,\xi')\\ &- \hat{q}_{-1}(t,x,\tau,\xi')\tilde{c}^{(o)}(t,x,\tau,\xi')\} = g_{-1}(t,x,\tau,\xi')\\ &+ \hat{q}_o(t,x,\tau,\xi')\hat{c}^{(-1)}(t,x,\tau,\xi')\,,\end{aligned} \tag{2.50}$$

on some $\Lambda^{\varepsilon'}_{\rho_o}$, $0 < \varepsilon' \leq \varepsilon$. Define:

$$
(2.51)\quad \begin{cases}
q_{-1}(t,x,\tau,\xi) = |\xi|^{-1}\, \hat{q}_{-1}(t,x,\frac{\tau}{|\xi|},\frac{\xi}{|\xi|}), \\
\tilde{e}^{(-1)}_m(x,\xi,t\tau) = |\xi|^{-m-1}\, \hat{e}^{(-1)}_m(x,\frac{\xi}{|\xi|},\frac{t\tau}{|\xi|}), \qquad m = 1,\dots,k-1 \\
\tilde{d}^{(-1)}_m(x,\xi,t\tau) = \hat{d}^{(-1)}_m(x,\frac{\xi}{|\xi|},\frac{t\tau}{|\xi|})\,|\xi|^{-1}, \\
\tilde{c}^{(-1)}_o(x,\xi,t\tau) = |\xi|^{-1}\, \hat{c}^{(-1)}_o(x,\frac{\xi}{|\xi|},\frac{t\tau}{|\xi|}), \\
\tilde{c}^{(-1)} = \sum_{m=1}^{k-1} \tilde{e}^{(-1)}_m \tau^m + \tilde{c}^{(-1)}_o + \sum_{m=1}^{k-1} \tilde{d}^{(-1)}_m t^m .
\end{cases}
$$

Modifying q_{-1} and $\tilde{c}^{(-1)}$ out of $V^{2\varepsilon'}_{\rho_o}$ we can suppose that q_{-1} and $\tilde{c}^{(-1)}$ are smoothly defined on $\dot{T}^*\tilde{X}$ (retaining the matrix structure). From (2.50), (2.51) it follows that equation (2.49) is satisfied on $V^{\varepsilon'}_{\rho_o}$. Therefore we choose Q and $\tilde{C}$ such that $Q - q_o(t,x,D_t,D_x) - q_{-1}(t,x,D_t,D_x)$ and $\tilde{C} - \tilde{c}^{(o)}(t,x,D_t,D_x) - \tilde{c}^{(-1)}(t,x,D_t,D_x)$ are in $L^{-2}(\tilde{X};\ N \times N)$.

Going on we obtain transport equations for q_{-2}, $\tilde{c}^{(-2)}$ which can be solved using Lemma 2.4 in the same set $\Lambda^{\varepsilon'}_{\rho_o}$, and so on. Defining $Q \sim \sum_{j\geq o} q_{-j}(t,x,D_t,D_x)$ and $\tilde{C} \sim \sum_{j\geq o} c^{(-j)}(t,x,D_t,D_x)$ we are done. The last assertion in the statement of Proposition 2.2 follows from the above proof taking into account Remark 2) after Lemma 2.4. q.e.d.

Propositions 2.1 and 2.2 give the following result.

THEOREM 2.1. Consider the Fuchsian system (2.1) and suppose that the Fuchs condition $(F)_{\rho_o}$ is satisfied at the point $\rho_o = (0,x_o,0,\xi^{(o)}) \in \Sigma_o$. Suppose that

$\rho_o \notin WF(f = Pu)$. Then:

(1) If for every $j \in \{1,2\}$ and for some choice of the sign + or - we have

(2.52) $$(\gamma_j^{\pm}(\rho_o) \setminus \{\rho_o\}) \cap WF(u) \cap V_{\rho_o} = \emptyset ,$$

for some conic neighborhood V_{ρ_o} of ρ_o, then $\rho_o \notin WF(u)$.

(2) i) If $\sigma(a_o(\rho_o)) \cap \{0,1,\dots\} = \emptyset$ and if

(2.53) $$[(\gamma_1^+(\rho_o) \cup \gamma_1^-(\rho_o)) \setminus \{\rho_o\}] \cap WF(u) \cap V_{\rho_o} = \emptyset ,$$

for some conic neighborhood V_{ρ_o} of ρ_o, then $\rho_o \notin WF(u)$.

ii) If $\#[\sigma(a_o(\rho_o)) \cap \{0,1,\dots\}] = h$ and if (2.53) is satisfied, then there exist distributions $v_1,\dots,v_h \in \mathcal{D}'(\tilde{X})^N$ with $WF(v_j) \subset \Sigma_2$ $j = 1,\dots,h$ such that $\rho_o \notin WF(u - \sum_{j=1}^{h} v_j)$.

(3) i) If $\sigma(a_o(\rho_o)) \cap \{-1, -2,\dots\} = \emptyset$ and if

(2.54) $$[(\gamma_2^+(\rho_o) \cup \gamma_2^-(\rho_o)) \setminus \{\rho_o\}] \cap WF(u) \cap V_{\rho_o} = \emptyset ,$$

for some conic neighborhood V_{ρ_o} of ρ_o, then $\rho_o \notin WF(u)$.

ii) If $\#[\sigma(a_o(\rho_o)) \cap \{-1, -2,\dots\}] = h$ and if (2.54) is satisfied, then there exist distributions $w_1,\dots,w_h \in \mathcal{D}'(\tilde{X})^N$ with $WF(w_j) \subset \Sigma_1$, $j = 1,\dots,h$ such that $\rho_o \notin WF(u - \sum_{j=1}^{h} w_j)$.

Proof. By Propositions 2.1 and 2.2, there exist p d o's $Q_1, Q_2 \in L^0(\tilde{X}; N \times N)$ elliptic near ρ_0 and a p d o $B(x,D_x) \in L^0(X;N)$ such that

$$(2.55) \qquad (t\,\partial_t I_N - A(t,x,D_t,D_x))Q_1 \equiv Q_2(t\,\partial_t I_N - B(x,D_x))$$

in a conic neighborhood of ρ_0 ; moreover, $b_0(x_0,\xi^{(0)}) = a_0(\rho_0)$, $b_0(x,\xi)$ being the principal symbol of the operator B.

Using (2.55) and the ellipticity of Q_1, Q_2 near ρ_0 we are reduced to prove the theorem for the system $t\,\partial_t I_N - B(x,D_x)$. Now all the assertions easily follow from theorems 1.1, 1.2 and the Corollary to Theorem 1.2.

q.e.d.

The next step consists in studying system (2.1) when the Fuchs condition $(F)_{\rho_0}$ is violated. Again, by Propositions 1.1 and 1.2 we are allowed to treat the case of a system (2.1) where the operator $A(t,x,D_t,D_x)$ has the following matrix structure:

$$(2.56) \qquad A(t,x,D_t,D_x) = \begin{pmatrix} B_{11}(x,D_x) & B_{12}(x,D_x)t & \cdots\cdots & B_{1k}(x,D_x)t^{k-1} \\ B_{21}(x,D_x)\partial_t & B_{22}(x,D_x) & \cdots\cdots & B_{2k}(x,D_x)t^{k-2} \\ B_{31}(x,D_x)\partial_t^2 & B_{32}(x,D_x)\partial_t & \cdots\cdots & B_{3k}(x,D_x)t^{k-3} \\ \cdots & \cdots & \cdots & \cdots \\ B_{k1}(x,D_x)\partial_t^{k-1} & B_{k2}(x,D_x)\partial_t^{k-2} & \cdots & B_{kk}(x,D_x) \end{pmatrix},$$

where:

(i) For $i = 1,\dots,k$, $B_{ii}(x,D_x) \in L^0(X; N_i \times N_i)$, $N_1 + N_2 + \dots N_k = N$, $k \geq 2$;

the principal symbol $b_{ii}^{(o)}(x,\xi)$ of the operator B_{ii} is such that $b_{ii}^{(o)}(x_o, \xi^{(o)})$ is in Jordan canonical form and has its spectrum reduced to the eigenvalue $\lambda - i + 1$, for some $\lambda \in \mathbb{C}$.

Moreover,

$$a_o(\rho_o) = \begin{pmatrix} b_{11}^{(o)}(x_o,\xi^{(o)}) & & & \square \\ & b_{22}^{(o)}(x_o,\xi^{(o)}) & & \\ & & \ddots & \\ \square & & & b_{kk}^{(o)}(x_o,\xi^{(o)}) \end{pmatrix}$$

(ii) $B_{ij}(x,D_x) \in L^{o}(X;\ N_i \times N_j)$ for $i, j = 1,\dots,k$, $i < j$;

$b_{ij}(x,D_x) \in L^{-(i-j)}(X;\ N_i \times N_j)$ for $i, j = 1,\dots,k$, $i > j$.

For the system $(t\,\partial_t I_N - A)u = f$ we write $u = (u_1,\dots,u_k)$, $f = (f_1,\dots,f_k)$ with $u_j, f_j \in \mathcal{D}'(\tilde{X})^{N_j}$, $j = 1,\dots,k$.

The following Lemma holds.

LEMMA 2.5. Let $u_j, f_j \in \mathcal{D}'(\tilde{X})^{N_j}$, $j = 1,\dots,k$, $k \geq 2$, be distributions for which $(t\,\partial_t I_N - A(t,x,D_t,D_x))\,(u_1,\dots,u_k) = (f_1,\dots,f_k)$, where A has the structure (2.56). Define

$$(2.57)\qquad \begin{cases} w_1 = (u_1,\ tu_2,\ t^2u_3,\dots,t^{k-1}u_k) \\ w_2 = (u_2,\ tu_3,\ t^2u_4,\dots,t^{k-2}u_k) \\ \dots\dots\dots\dots\dots\dots\dots\dots \\ w_{k-1} = (u_{k-1},\ tu_k)\ , \\ w_k = u_k . \end{cases}$$

Then the vector $w = (w_1,\dots,w_k)$ satisfies the system

$$(2.58)\qquad (t\,\partial_t\, I_{\frac{k(k+1)}{2}N} - \tilde{A}(t,x,D_t,D_x))w = g$$

where $g \in \mathcal{D}'(\tilde{X})^{\frac{k(k+1)}{2}N}$ and $WF(g) = WF(f)$, while $\tilde{A} \in L^0(\tilde{X};\ \frac{k(k+1)}{2}N \times \frac{k(k+1)}{2}N\ )$ has the following matrix structure.

$$(2.59)\qquad \tilde{A}(t,x,D_t,D_x) = \begin{pmatrix} \tilde{B}_{11}(t,x,D_t,D_x) & & \square \\ \tilde{B}_{21}(t,x,D_t,D_x) & \tilde{B}_{22}(t,x,D_t,D_x) & \\ \dots\dots\dots\dots & \dots\dots\dots\dots & \dots\dots \\ \tilde{B}_{k1}(t,x,D_t,D_x) & \tilde{B}_{k2}(t,x,D_t,D_x) & \dots\tilde{B}_{kk}(t,x,D_t,D_x) \end{pmatrix}$$

with

i) $\tilde{B}_{ii} \in L^{0}(\tilde{X};\ (N_i+\ldots+N_k) \times (N_i+\ldots+N_k))$; the principal symbol $b_{ii}^{(o)}(t,x,\tau,\xi)$ of the operator B_{ii} is such that the spectrum of $b_{ii}^{(o)}(\rho_o)$ is reduced to the eigenvalue $\lambda - i + 1$, $i = 1,\ldots,k$.

ii) $\tilde{B}_{ij} \in L^{0}(\tilde{X};\ (N_i+\ldots+N_k) \times (N_j+\ldots+N_k))$ for $i, j = 1,\ldots,k$, $i > j$; moreover, the principal symbol of $\tilde{B}_{ij}$ vanishes at ρ_o.

Proof. We start by defining $w = tu_k$ and find what system is satisfied by the vector $(u_1, u_2,\ldots,u_{k-2}, (u_{k-1}, tu_k), u_k)$.

For $j = 1,\ldots,k$, from (2.56) we obtain

$$t\,\partial_t u_j = \sum_{\ell=1}^{j} B_{j\ell}(x,D_x)\,\partial_t^{j-\ell}\, u_\ell + \sum_{\ell=j+1}^{k} B_{j\ell}(x,D_x)\, t^{\ell-j}\, u_\ell + f_j .$$

Now

$$t\,\partial_t w = t\,\partial_t (t\,u_k) = t\,u_k + t(t\,\partial_t u_k) = \sum_{\ell=1}^{k-1} B_{k\ell}(x,D_x)\partial_t^{k-\ell}\, u_\ell +$$

$$+ (I_{N_k} + B_{kk}(x,D_x))w + t\,f_k .$$

It follows that the vector $(u_1,\ldots,u_{k-2}, (u_{k-1}, t\,u_k), u_k)$ satisfies the system

$$(2.60)\quad t\,\partial_t \begin{pmatrix} u_1 \\ \vdots \\ \begin{pmatrix} u_{k-1} \\ t\,u_k \end{pmatrix} \\ u_k \end{pmatrix} = \begin{pmatrix} B_{11} & B_{12}\,t & \cdots & \tilde{B}_{1,k-1}\,t^{k-2} & 0 \\ & \ddots & & & \\ \tilde{B}_{k-1,1} & \tilde{B}_{k-1,2} & \cdots & \tilde{B}_{k-1,k-1} & 0 \\ B_{k1}\,\partial_t^{k-1} & B_{k2}\,\partial_t^{k-2} & \cdots & \tilde{B}_{k,k-1} & B_{kk} \end{pmatrix} + \begin{pmatrix} f_1 \\ \vdots \\ \begin{pmatrix} f_{k-1} \\ t\,f_k \end{pmatrix} \\ f_k \end{pmatrix}$$

where $\tilde{B}_{1,k-1} = (B_{1,k-1}\;B_{1k}),\ldots,\tilde{B}_{k-2,k-1} = (B_{k-2,k-1}\;B_{k-2,k})$,

$$\tilde{B}_{k-1,1} = \begin{pmatrix} B_{k-1,1}\,\partial_t^{k-2} \\ B_{k-1}\,t\,\partial_t^{k-1} \end{pmatrix},\quad \tilde{B}_{k-1,2} = \begin{pmatrix} B_{k-1,2}\,\partial_t^{k-3} \\ B_{k,2}\,t\,\partial_t^{k-2} \end{pmatrix},\quad \ldots,$$

$$\tilde{B}_{k-1,k-1} = \begin{pmatrix} B_{k-1,k-1} & B_{k-1,k} \\ B_{k,k-1}\,t\,\partial_t & I + B_{kk} \end{pmatrix},\quad \tilde{B}_{k,k-1} = (B_{k,k-1}\,\partial_t\quad 0)\ .$$

Note that the spectrum of the principal symbol of $\tilde{B}_{k-1,k-1}$ at ρ_0 is reduced to the eigenvalue $\lambda - k + 2$ and that all the operators involved in the r.h.s. of (2.60) have order zero. Note that $WF(u_1,\ldots,u_{k-2},(u_{k-1},t\,u_k),u_k)) = WF(u)$ and $WF(f_1,\ldots,f_{k-2},(f_{k-1},\,t\,f_k),f_k) = WF(f)$.

If we put $w = t(u_{k-1},\,t\,u_k)$ and proceed as above we see that the vector

$(u_1,u_2,\ldots,(u_{k-2},t(u_{k-1},t\,u_k)),\ (u_{k-1},t\,u_k),u_k)$ satisfies a system of the same type where the two last columns are cleaned up and with new blocks on the diagonal $B_{11},\ldots,\tilde{B}_{k-2,k-2},\ \tilde{B}_{k-1,k-1},\ B_{kk}$, where the spectrum of the principal symbol of $\tilde{B}_{k-2,k-2}$ at ρ_0 is reduced to the eigenvalue $\lambda - k + 3$ and all the operators in the matrix are of order zero. Proceeding in this way we arrive at the form (2.59).

q.e.d.

We now use Proposition 2.2 once more, taking into account Remark 1 after Lemma 2.4. Precisely, considering the system $(t\,\partial_t - \tilde{A}(t,x,D_t,D_x))\ w = g$, where $\tilde{A}$ is given in (2.59), we deduce that there exist two p d o's $Q_1,\ Q_2 \in L^0\ (\tilde{X};\ \frac{k(k+1)}{2}N \times \frac{k(k+1)}{2}N)$ elliptic near ρ_0 such that in a conic neighborhood of ρ_0 we have $(t\,\partial_t - \tilde{A})Q_1 \equiv Q_2(t\,\partial_t - \hat{A})$, where $\hat{A} \in L^0(\tilde{X};\frac{k(k+1)}{2}N \times \frac{k(k+1)}{2}N)$ has the following matrix structure

$$(2.61)\quad \hat{A} = \begin{pmatrix} \hat{B}_{11}(x,D_x) & & & \boxed{} \\ \hat{B}_{21}(x,D_x)\partial_t & \hat{B}_{22}(x,D_x) & & \\ \cdots & \cdots & \cdots & \cdots \\ \hat{B}_{k1}(x,D_x)\partial_t^{k-1} & \hat{B}_{k2}(x,D_x)\partial_t^{k-2} & \cdots\cdots & \hat{B}_{kk}(x,D_x) \end{pmatrix}$$

with:

i) $\hat{B}_{ii} \in L^O(X;(N_i+\ldots+N_k) \times (N_i+\ldots+N_k))$ and the spectrum of the principal symbol of $\hat{B}_{ii}$ at $(x_0,\xi^{(o)})$ consists of the eigenvalue $\lambda - i + 1$, $i = 1,\ldots,k$.

ii) $\hat{B}_{ij} \in L^{-(i-j)}(X;(N_i+\ldots+N_k) \times (N_j+\ldots+N_k))$ for $i,j = 1,\ldots,k$, $i > j$.

Now consider the system $(t\,\partial_t I_{\frac{k(k+1)}{2}N} - \hat{A})\hat{w} = \hat{g}$, and define

$$
(2.62)\qquad \begin{cases}
\tilde{w}_1 = (\hat{w}_1, \partial_t\hat{w}_1, \dots, \partial_t^{k-1}\hat{w}_1) \\
\tilde{w}_2 = (\hat{w}_2, \partial_t\hat{w}_2, \dots, \partial_t^{k-2}\hat{w}_2) \\
\dots\dots\dots\dots\dots\dots\dots\dots \\
\tilde{w}_{k-1} = (\hat{w}_{k-1}, \partial_t\hat{w}_{k-1}) \\
\tilde{w}_k = \hat{w}_k
\end{cases}
$$

Then it is easy to recognize that the vector $(\tilde{w}_1,\dots,\tilde{w}_k) = \tilde{w}$ satisfies a system:

$$
(2.63)\qquad (t\,\partial_t I_M - B(x,D_x))\tilde{w} = \tilde{g}\ , \quad M = \left(\frac{k(k+1)}{2}\right)^2 N\ ,
$$

where $\tilde{g} \in \mathcal{D}'(\tilde{X})^M$ with $WF(\tilde{g}) = WF(\hat{g})$ and $B(x,D_x) \in L^0(X;\ M\times M)$ is such that the spectrum of its principal symbol at ρ_0 consists of the eigenvalues $\lambda, \lambda-1,\dots,\lambda-k+1$. Note that $WF(\tilde{w}) = WF(\hat{w})$.

We can now state one of the conclusive results of this Section.

THEOREM 2.2 For a general Fuchsian system (2.1) the conclusions (1), (2) and (3) of Theorem 2.1 hold.

<u>Proof.</u> Using Proposition 1.2 we are reduced to consider a finite number of decoupled systems $(t\,\partial_t I_{N_i} - A_i(t,x,D_t,D_x))v_i = g_i$ $i=1,\dots,k$. If $a_i^{(o)}(\rho_0)$ satisfies the Fuchs condition $(F)_{\rho_0}$ ($a_i^{(o)}$ being the principal symbol of the operator A_i), Theorem 2.1 can be applied to the component v_i. For each of those blocks A_i for which $(F)_{\rho_0}$ is not satisfied we apply the above procedure in order to deduce a big

system $t\partial_t I_{M_i} - B_i(x,D_x))\tilde{w}_i = \tilde{g}_i$ with $WF(\tilde{w}_i) = WF(v_i)$ and $WF(\tilde{g}_i) = WF(g_i)$ near ρ_o, so that we can use the results of Section 1 to deduce the properties of $WF(v_i)$ near ρ_o.

q.e.d.

To complete the microlocal analysis of a general Fuchsian system (2.1):

$$Pu = t\partial_t I_N u - A(t,x,D_t,D_x)u = f \,, \tag{2.64}$$

we now turn to the study of the singularities of u near the conormal bundle $\dot{N}^*X = N^*X \setminus \tilde{X}$ of X.

Define $N^*_\pm X = \{(t,x,\tau,\xi) \in \dot{T}^*\tilde{X} \mid t = 0, \xi = 0, \ \pm\tau > 0\}$ so that $\dot{N}^*X = N^*_+X \cup N^*_-X$.

Denote by $\mathcal{M}(\tilde{X})$ the sheaf of microdistributions on $\tilde{X}$ and given $\rho_o \in \dot{N}^*X$, let $\mathcal{M}_{\rho_o}(\tilde{X})$ be the stalk of $\mathcal{M}(\tilde{X})$ over ρ_o. The operator P defines a linear map $P : \mathcal{M}_{\rho_o}(\tilde{X})^N \longrightarrow \mathcal{M}_{\rho_o}(\tilde{X})^N$. We are interested in finding ker P and Coker P. In the scalar case $N = 1$, N. Hanges [12, Proposition 2.7] reduced the study of the map $P : \mathcal{M}_{\rho_o}(\tilde{X}) \longrightarrow \mathcal{M}_{\rho_o}(\tilde{X})$ to that of the map $t : \mathcal{M}_{\rho_o}(\tilde{X}) \longrightarrow \mathcal{M}_{\rho_o}(\tilde{X})$, which is quite simple.

In the vector-valued case the situation is more complicated and we find once more the obstruction given by the *Fuchs condition* :

$$(F)_{\rho_o} : \qquad \sigma(a_o(\rho_o) + j\, I_N) \cap \sigma(a_o(\rho_o)) = \emptyset \,, \quad \forall j \in \mathbb{Z} \setminus \{0\} \,,$$

with, this time, $\rho_o \in \dot{N}^*X$.

To bypass the Fuchs condition we proceed as in the case $\rho_o \in \Sigma_o$ treated above (but with some important differences). The following treatment has been greatly influenced by the ideas of Kashiwara - Oshima [19] (see also Tahara [28]).

First we use a trick of Chazarain [9] to show that near $\dot{N}^*X$ we can get rid of the dependence on t in the operator $A(t,x,D_t,D_x) \in L^0(\tilde{X}; N \times N)$.

LEMMA 2.6. There exist p d o's $E, \tilde{A} \in L^0(\tilde{X}; N \times N)$ such that :

1) E is elliptic near $\dot{N}^*X$.

2) $\tilde{A} = \tilde{A}(x,D_t,D_x)$, i.e. the symbol of the operator $\tilde{A}$ is independent of t. Moreover, the principal symbol $\tilde{a}_0$ of $\tilde{A}$ satisfies $\tilde{a}_0(x,\tau,\xi) = a_0(0,x,\tau,\xi)$.

3) $(t\,\partial_t I_N - A)E \equiv t\,\partial_t I_N - \tilde{A}$, near $\dot{N}^*X$.

Proof. Since $\partial_t I_N$ is elliptic near $\dot{N}^*X$, denoting by $D_t^{-1} \in L^{-1}(\tilde{X})$ a p.d.o. with symbol $\frac{1}{\tau}$ in a conic neighborhood of $\dot{N}^*X$, we have $(D_tI_N)(D_t^{-1}I_N) \equiv (D_t^{-1}I_N)(D_tI_N) \equiv I_N$ near $\dot{N}^*X$.

Application of $D_t^{-1}I_N$ to the system $t\,\partial_t I_N - A(t,x,D_t,D_x)$ yields an operator

$$t\, I_N - B(t,x,D_t,D_x)\ , \tag{2.65}$$

where $B = \frac{1}{i} A\, D_t^{-1} \in L^{-1}(\tilde{X}; N \times N)$ and $b_{-1}(t,x,\tau,\xi) = \frac{1}{i\tau}\, a_0(t,x,\tau,\xi)$ near $\dot{N}^*X$ (b_{-1} being the principal symbol of the operator B).

Now we look for operators $F \sim I_N + F_{-1} + \ldots$, $F_{-j} \in L^{-j}(\tilde{X}; N \times N)$, $C \sim C_{-1} + C_{-2} + \ldots$, $C_{-j} \in L^{-j}(\tilde{X}; N \times N)$, $j \geq 1$, with the C_{-j} independent of t, such that

$$(t\, I_N - B)F \equiv t\, I_N - C\ , \quad \text{near } \dot{N}^*X\ . \tag{2.66}$$

First, try to find F_{-1} and C_{-1} so that

(2.67) $$(t\, I_N - B)(I_N + F_{-1}) \equiv t\, I_N - C_{-1}\ , \quad \text{mod}\ L^{-2}(\tilde{X};\ N \times N).$$

Define $c_{-1}(x,\tau,\xi) = b_{-1}(0,x,\tau,\xi)$, so that to implement (2.67) it is enough to choose

$$f_{-1}(t,x,\tau,\xi) = \frac{1}{t}\,[b_{-1}(t,x,\tau,\xi) - b_{-1}(0,x,\tau,\xi)] = \int_0^1 \frac{\partial b_{-1}}{\partial t}(\sigma t,x,\tau,\xi)\,d\sigma\ ,$$

which is a smooth symbol.

Suppose we have already found $F_{-1},\dots,F_{-k},\ C_{-1},\dots,C_{-k}$ so that :

(2.68) $$(t\, I_N - B)(I_N + F_{-1}+\dots+F_{-k}) \equiv t\, I_N - (C_{-1}+\dots+C_{-k}),\quad \text{mod}\ L^{-(k+1)}(\tilde{X};\ N \times N)\ ,$$

and look for $F_{-(k+1)}$, $C_{-(k+1)}$ in order that (2.68) is satisfied with k replaced by $k+1$. We must have

(2.69) $$(t\, I_N - B)F_{-(k+1)} = C_{-(k+1)} + G_{-(k+1)}\ ,$$

for some $G_{-(k+1)} \in L^{-(k+1)}(\tilde{X};\ N \times N)$, uniquely determined mod $L^{-(k+2)}$. Define $c_{-(k+1)}(x,\tau,\xi) = -\,g_{-(k+1)}(0,x,\tau,\xi)$, where $g_{-(k+1)}$ is the principal symbol of $G_{-(k+1)}$. Then choose

$$f_{-(k+1)}(t,x,\tau,\xi) = \int_0^1 \frac{\partial g_{-(k+1)}}{\partial t}(\sigma t,x,\tau,\xi)\,d\sigma.$$

Thus (2.66) is proved. Defining $\tilde{A}(x,D_t,D_x) = i\, C(x,D_t,D_x)D_t$ and $E = (i\, D_t)^{-1}$ $F(i\, D_t)$ we are finished.

q.e.d.

From now on we shall restrict to consider a Fuchsian system of the type:

$$(2.70) \qquad Pu = (t\,\partial_t I_N - A(x,D_t,D_x))u = f \;,$$

where $A \in L^{O}(\tilde{X};\, N \times N)$ is independent of t .

We will show that modulo elliptic operators, system (2.70) can be reduced to a canonical form near N_+^*X or N_-^*X . We prepare this reduction by proving some preliminary results.

LEMMA 2.7. Let M be a C^∞ countable manifold and let $a(z,\eta)$ be a smooth $N \times N$ matrix defined for $(z,\eta) \in M \times \mathbb{R}^n$.

Suppose that at some point $z_o \in M$ the matrix $a(z_o,0)$ is block diagonal: $a(z_o,0) = (a_{jj}(z_o))$, with blocks $a_{jj}(z_o)$ of dimension $N_j \times N_j$, $j=1,\ldots,k$, $k \geqq 2$, $N_1+\ldots+N_k = N$.

Moreover, suppose that :

$$(2.71) \qquad \sigma(a_{jj}(z_o) + r\, I_{N_j}) \cap \sigma(a_{ii}(z_o)) = \emptyset \;, \quad \begin{cases} i,j = 1,\ldots,k \;,\; i \neq j \;, \\ r \in \mathbb{Z} \;. \end{cases}$$

Then there is a neighborhood $U \times (B_\varepsilon = \{\eta \in \mathbb{R}^n \mid |\eta| < \varepsilon\})$ of $(z_o,0)$ for which the following holds:

1 - There are two $N \times N$ smooth matrices $e(z,\eta)$, $\tilde{a}(z,\eta)$ defined on $M \times \mathbb{R}^n$ such that

i) $e(z_0,0) = I_N$;

ii) $\tilde{a}(z,\eta) = (\tilde{a}_{jj}(z,\eta))_{j=1,\dots,k}$ is block diagonal with $\tilde{a}_{jj}(z_0,0) = a_{jj}(z_0)$ $j = 1,\dots,k$;

iii) Putting $\eta \cdot \nabla_\eta = \sum_{j=1}^{n} \eta_j \frac{\partial}{\partial \eta_j}$:

$$(\eta \cdot \nabla_\eta e)(z,\eta) - \{a(z,\eta)e(z,\eta) - e(z,\eta)\tilde{a}(z,\eta)\} = 0 , \tag{2.72}$$

for $(z,\eta) \in U \times B_\varepsilon$.

2 - For $h = 1,2,\dots,$ and for every smooth $N \times N$ matrix g_h, defined on $M \times \mathbb{R}^n$, there exist two smooth $N \times N$ matrices $e_{-h}(z,\eta)$, $\tilde{a}_{-h}(z,\eta)$ on $M \times \mathbb{R}^n$ such that:

i) $\tilde{a}_{-h}(z,\eta)$ is block diagonal with k blocks of dimension $N_j \times N_j$, $j = 1,\dots,k$;

ii)

$$\begin{aligned}(h\, I_N + \eta \cdot \nabla_\eta)e_{-h}(z,\eta) &- \{a(z,\eta)e_{-h}(z,\eta) - e_{-h}(z,\eta)\tilde{a}(z,\eta)\} \\ &+ e(z,\eta)\tilde{a}_{-h}(z,\eta) = g_h(z,\eta) , \quad \text{for } (z,\eta) \in U \times B_\varepsilon ,\end{aligned} \tag{2.72'}$$

where $\tilde{a}$, and e are the matrices constructed in 1- .

<u>Proof.</u> Without loss of generality we can assume $k = 2$; the general case follows by induction on k .

Consider the Taylor expansion $a(z,\eta)\sim\sum_\alpha a^{(\alpha)}(z)\eta^\alpha$ and seek formal power series $\sum_\alpha e^{(\alpha)}(z)\eta^\alpha$, $\sum_\alpha \tilde{a}^{(\alpha)}\eta^\alpha$ such that (2.72) is satisfied at the formal level. Observing that $\eta\cdot\nabla_\eta\eta^\alpha=|\alpha|\eta^\alpha$, to solve (2.72) we are led to the following equations:

$$|\alpha|e^{(\alpha)}(z)-\left\{\sum_{\gamma\leqq\alpha}a^{(\alpha-\gamma)}(z)e^{(\gamma)}(z)-\sum_{\gamma\leqq\alpha}e^{(\alpha-\gamma)}(z)\tilde{a}^{(\gamma)}(z)\right\}=0\,, \tag{2.73_α}$$

$$\alpha\in\mathbb{Z}^n_+ .$$

For $\alpha=0$ we have :

$$a^{(o)}(z)e^{(o)}(z)-e^{(o)}(z)\tilde{a}^{(o)}(z)=0\,, \tag{2.73_o}$$

with $e^{(0)}(z_0)=I_N$ and $\tilde{a}^{(0)}(z_0)=a^{(0)}(z_0)=a(z_0,0)$.

As in the proof of Lemma 2.1, there is a neighborhood W of z_0 in M and smooth matrices

$$e^{(0)}(z)=\begin{pmatrix} I_{N_1} & e^{(0)}_{12}(z)\\ e^{(0)}_{21}(z) & I_{N_2}\end{pmatrix},\qquad \tilde{a}^{(0)}(z)=\begin{pmatrix}\tilde{a}^{(0)}_{11}(z) & \bigcirc\\ \bigcirc & \tilde{a}^{(0)}_{22}(z)\end{pmatrix}$$

defined on M, for which equation $(2.73)_0$ is satisfied when $z\in W$.

Now we observe that with $\tilde{a}^{(0)}(z)$ and $e^{(0)}(z)$ determined as above, there is a fixed neighborhood W' of z_0, with $\overline{W'}\subset W$, such that for all $r\in\{0,1,2,..$ the linear maps which map

$$(q;\ \tilde{a}) = \begin{pmatrix} \tilde{a}_{11} & q_{12} \\ q_{21} & \tilde{a}_{22} \end{pmatrix} \quad \text{into}$$

$$(2.74) \qquad \mathscr{L}_r(q;\tilde{a}) = r\, I_N(q;0) - \{a^{(o)}(z)(q;0) - (q;0)\tilde{a}^{(o)}(z)\} + e^{(o)}(z)(0;\tilde{a}) ,$$

are smoothly invertible on W'.

Suppose we have already constructed $(e^{(\beta)}(z);\ \tilde{a}^{(\beta)}(z))$ for $\beta < \alpha$ such that $(2.73)_\beta$ is satisfied on W'. To construct $(e^{(\alpha)}(z);\ \tilde{a}^{(\alpha)}(z))$ we rewrite equation $(2.73)_\alpha$ as :

$$(2.73)'_\alpha \qquad |\alpha|\, e^{(\alpha)}(z) - \{a^{(o)}(z)e^{(\alpha)}(z) - e^{(\alpha)}(z)\tilde{a}^{(o)}(z)\} + e^{(o)}(z)\tilde{a}^{(\alpha)}(z) = g_\alpha(z) ,$$

where $g_\alpha(z)$ is a known smooth matrix defined on W'.

Using (2.74) with $r = |\alpha|$, we find $e^{(\alpha)}(z) = (q;0), \tilde{a}^{(\alpha)}(z) = (0;\tilde{a})$ as smooth matrices on W' for which equation $(2.73)'_\alpha$ is satisfied in W'.

Having constructed $e^{(\alpha)}(z), \tilde{a}^{(\alpha)}(z)$ for $\alpha \geqq 0$ and $z \in W'$ we smoothly modify these matrices outside of a fixed neighborhood V of z_o, $\bar{V} \subset W'$, so that we can suppose that they are defined on M.

Using Borel's Lemma we find two smooth $N \times N$ matrices $\tilde{e}(z,\eta), \tilde{a}(z,\eta)$ defined on $M \times \mathbb{R}^n$ such that $\tilde{e}(z,\eta) \sim \sum_\alpha e^{(\alpha)}(z)\eta^\alpha$, $\tilde{a}(z,\eta) \sim \sum_\alpha a^{(\alpha)}(z)\eta^\alpha$ and :

$$(\eta \cdot \nabla_\eta\, \tilde{e}) - \{a\, \tilde{e} - \tilde{e}\, \tilde{a}\} = \psi, \quad \text{on} \quad V \times \mathbb{R}^n ,$$

where $\psi(z,\eta)$ is a smooth matrix which is flat at $\eta = 0$, $z \in V$ (i.e. $(\partial_\eta^\alpha \psi)(z,0) = 0,\ \forall\, \alpha \in \mathbf{Z}_+^n,\ \forall\, z \in V$).

Using polar coordinates in $\mathbb{R}^n$, write $\eta = r\omega$, $r \geq 0$, $\omega \in S^{n-1}$, and define $A(z,\omega;r) = a(z,r\omega)$, $\tilde{A}(z,\omega;r) = \tilde{a}(z,r\omega)$, $\Psi(z,\omega;r) = \psi(z;r\omega)$. Consider the equation:

$$(2.75) \qquad r\frac{d}{dr} f(z,\omega;r) - \{A(z,\omega;r)f(z,\omega;r) - f(z,\omega;r)\tilde{A}(z,\omega;r)\} = -\Psi(z,\omega;r) .$$

Using Lemma 2.0 (or rather its proof), we can find a smooth $N \times N$ matrix $f(z,\omega;r)$ defined on $V \times S^{n-1} \times [0,+\infty)$, which is flat at $r = 0$ and satisfies (2.75) for $(z,\omega) \in V \times S^{n-1}$ and $r \in [0,\varepsilon')$, for some $\varepsilon' > 0$. Defining $\tilde{\tilde{e}}(z,\eta) = f(z, \eta/|\eta| ; |\eta|)$, we see that $\tilde{\tilde{e}}$ is smoothly defined on $V \times \mathbb{R}^n$, flat at $\eta = 0$, and such that

$$(\eta \cdot \nabla_\eta \tilde{\tilde{e}}) - \{a \tilde{\tilde{e}} - \tilde{\tilde{e}} \tilde{a}\} = -\psi, \quad \text{for} \quad (z,\eta) \in V \times B_{\varepsilon'} .$$

Modifying $\tilde{\tilde{e}}(z,\eta)$ outside of a neighborhood $U \times B_\varepsilon$ of $(z_0,0)$, with $\bar{U} \subset V$, $0 < \varepsilon < \varepsilon'$, we can suppose that $\tilde{\tilde{e}}$ is defined on $M \times \mathbb{R}^n$. Putting $e(z,\eta) = \tilde{e}(z,\eta) + \tilde{\tilde{e}}(z,\eta)$ we see that part 1 - is proved.

To prove part 2 - we proceed as above by considering the formal power series $e_{-h}(z,\eta) \sim \sum_\alpha e_{-h}^{(\alpha)}(z)\eta^\alpha$, $\tilde{a}_{-h}(z,\eta) \sim \sum_\alpha a_{-h}^{(\alpha)}(z)\eta^\alpha$.

Again, let $(q;\tilde{a}) = \begin{pmatrix} \tilde{a}_{11} & q_{12} \\ q_{21} & \tilde{a}_{22} \end{pmatrix}$ and seek $e_{-h}^{(\alpha)}$ in the form $(q;0)$ and $\tilde{a}_{-h}^{(\alpha)}$ in the form $(0;\tilde{a})$. We are led to transport equations of the form :

$$(2.76)_\alpha \qquad \begin{aligned} &(h+|\alpha|)I_N(q;0) - \{a^{(o)}(z)(q;0) - (q;0)\tilde{a}^{(o)}(z)\} \\ &+ e^{(o)}(z)(0;\tilde{a}) = g_h^{(\alpha)}(z) + \text{known data.} \end{aligned}$$

Since $h+|\alpha| \geq 1$ for $|\alpha| \geq 0$, from (2.74) it follows that equation $(2.76)_\alpha$

are smoothly solvable on W'. From now on we proceed exactly as before.

q.e.d.

LEMMA 2.8. Let M be as in Lemma 2.7 and suppose we are given a smooth $N \times N$ matrix $a(z,\eta)$ defined on $M \times \mathbb{R}^n$ such that at some point $z_o \in M$ and for some $\lambda \in \mathbb{C}$ we have :

$a(z_o,0) = (a_{jj}(z_o))_{j=1,\ldots,\ell}$ is block diagonal with blocks a_{jj} of dimension $N_j \times N_j$, $N_1 + N_2 + \ldots + N_\ell = N$, for which

1) $a_{11}(z_o)$ is upper triangular

2) $\sigma(a_{jj}(z_o)) = \{\lambda - k_j\}$,

where $0 = k_1 < k_2 < \ldots < k_\ell$ are some non-negative integers.

Then there is a neighborhood $U \times B_\varepsilon$ of $(z_o,0)$ for which the following holds:

1 - There are two smooth $N \times N$ matrices $q(z,\eta)$, $c(\hat{z},\eta)$ defined on $M \times \mathbb{R}^n$ such that:

i) $q(z_o,0) = I_N$;

ii) $c(z,\eta) = (c_{ij}(z,\eta))_{i,j=1,\ldots,\ell}$, where c_{ij} is a $N_i \times N_j$ matrix, with:

a) $c_{ij}(z,\eta) = 0$ if $i > j$,

b) $c_{jj}(z,\eta) = c_{jj}(z)$ is independent of η and $c_{jj}(z_o) = a_{jj}(z_o)$

c) $c_{ij}(z,\eta) = \sum_{|\alpha| = k_j - k_i} c_{ij}^{(\alpha)}(z)\eta^\alpha$ for $i < j$, for some smooth matrices $c_{ij}^{(\alpha)}(z)$.

iii)

(2.77) $$(\eta \cdot \nabla_\eta q)(z,\eta) - \{a(z,\eta)q(z,\eta) - q(z,\eta)c(z,\eta)\} = 0 ,$$

for $(z,\eta) \in U \times B_\varepsilon$.

2 - For $h = 1,2,\ldots,$ and for every smooth $N \times N$ matrix g_h, defined on $M \times \mathbb{R}^n$, there exist two smooth matrices $q_{-h}(z,\eta)$, $c_{-h}(z,\eta)$ defined on $M \times \mathbb{R}^n$, such that:

i) $c_{-h}(z,\eta) = (c_{ij,-h}(z,\eta))_{i,j=1,\ldots,\ell}$, where $c_{ij,-h}$ is a $N_i \times N_j$ matrix with:

a) $c_{ij,-h}(z,\eta) = 0$ for $i > j - h$,

b) $c_{ij,-h} = \sum_{|\alpha| = k_j - k_i - h} c_{ij,-h}^{(\alpha)}(z)\eta^\alpha$ for $i \leq j - h$, for some smooth matrices $c_{ij,-h}^{(\alpha)}(z)$

(in particular, if $h \geq \ell$, $c_{-h}(z,\eta) \equiv 0$);

ii)

(2.77)' $$(h I_N + \eta \cdot \nabla_\eta)q_{-h}(z,\eta) - \{a(z,\eta)q_{-h}(z,\eta) - q_{-h}(z,\eta)c(z,\eta)\} + q(z,\eta)c_{-h}(z,\eta) = g_h(z,\eta), \text{ for } (z,\eta) \in U \times B_\varepsilon,$$

where q and c are the matrices constructed in 1 -.

Proof. For the sake of simplicity we treat the case $\ell = 2$, $k_1 = 0$, $k_2 = 1$, leaving to the reader to supply for the argument in the general case. First we try to solve (2.77) at the formal power series level. Writing $a(z,\eta) \sim \sum_\alpha a^{(\alpha)}(z)\eta^\alpha$, we look for formal series $\sum_{\alpha} q^{(\alpha)}(z)\eta^{\alpha}$, $\sum_{\alpha} c^{(\alpha)}(z)\eta^{\alpha}$, with

$$q^{(o)}(z) = \begin{pmatrix} I_{N_1} & q_{12}^{(o)}(z) \\ q_{21}^{(o)}(z) & I_{N_2} \end{pmatrix}, \quad c^{(o)}(z) = \begin{pmatrix} c_{11}^{(o)}(z) & \square \\ \square & c_{22}^{(o)}(z) \end{pmatrix},$$

$$q^{(\alpha)}(z) = \begin{pmatrix} q_{11}^{(\alpha)}(z) & \square \\ q_{21}^{(\alpha)}(z) & q_{22}^{(\alpha)}(z) \end{pmatrix}, \quad c^{(\alpha)}(z) = \begin{pmatrix} \square & c_{12}^{(\alpha)}(z) \\ \square & \square \end{pmatrix}, \text{ for } |\alpha| = 1,$$

$$q^{(\alpha)}(z) = \begin{pmatrix} q_{11}^{(\alpha)}(z) & q_{12}^{(\alpha)}(z) \\ q_{21}^{(\alpha)}(z) & q_{22}^{(\alpha)}(z) \end{pmatrix}, \quad c^{(\alpha)}(z) = \square \quad , \text{ for } |\alpha| > 1.$$

We are led to the following transport equations :

$$(2.78)_o \qquad a^{(o)}(z)q^{(o)}(z) - q^{(o)}(z)c^{(o)}(z) = 0 .$$

$$(2.78)_1 \qquad \begin{aligned} &I_N\, q^{(\alpha)}(z) - \{a^{(o)}(z)q^{(\alpha)}(z) - q^{(\alpha)}(z)c^{(o)}(z)\} + \\ &q^{(o)}(z)c^{(\alpha)}(z) = a^{(\alpha)}(z)q^{(o)}(z) , \quad \text{for } |\alpha| = 1. \end{aligned}$$

$$(2.78)_\alpha \qquad \begin{aligned} &|\alpha| I_N\, q^{(\alpha)}(z) - \{a^{(o)}(z)q^{(\alpha)}(z) - q^{(\alpha)}(z)c^{(o)}(z)\} = \\ &= \sum_{0<\beta\leqq\alpha} a^{(\beta)}(z)q^{(\alpha-\beta)}(z) - \sum_{0<\beta\leqq\alpha} q^{(\alpha-\beta)}(z)c^{(\beta)}(z), \quad \text{for } |\alpha| > 1. \end{aligned}$$

Equation $(2.78)_o$ is smoothly solved in a neighborhood of z_o using the implicit function theorem as in Lemma 2.1. The linear equation $(2.78)_1$ and $(2.78)_\alpha$ are smoothly solvable in a fixed neighborhood of z_0 .

Then, there exists a smooth $N \times N$ matrix

$$(2.79) \quad c(z,\eta) = \begin{pmatrix} c_{11}^{(0)}(z) & \sum_{|\alpha|=1} c_{12}^{(\alpha)}(z)\eta^{\alpha} \\ \square & c_{22}^{(0)}(z) \end{pmatrix}$$

defined on $M \times \mathbb{R}^n$ and satisfying $c_{jj}^{(0)}(z_0) = a_{jj}(z_0)$, $j = 1,2$.

Using Borel's Lemma we construct a smooth $\tilde{q}(z,\eta) \sim \sum_{\alpha} q^{(\alpha)}(z)\eta^{\alpha}$ defined on $M \times \mathbb{R}^n$ such that $q(z_0,0) = I_N$ and :

$$(2.80) \qquad (\eta \cdot \nabla_{\eta} \tilde{q})(z,\eta) - \{a(z,\eta)\tilde{q}(z,\eta) - \tilde{q}(z,\eta)c(z,\eta)\} = \psi(z,\eta) \text{ , on } M \times \mathbb{R}^n$$

for some matrix $\psi(z,\eta)$ which is flat at $\eta = 0$ on some neighborhood V of z_0.

Using the same argument of Lemma 2.7, we can add to $\tilde{q}(z,\eta)$ a smooth matrix $\tilde{\tilde{q}}(z,\eta)$, defined on $M \times \mathbb{R}^n$, such that, putting $q(z,\eta) = \tilde{q}(z,\eta) + \tilde{\tilde{q}}(z,\eta)$, equation (2.77) is satisfied in a neighborhood $U \times B_{\varepsilon}$ of $(z_0,0)$.

Part 1 - is proved.

Part 2 - is proved by a similar argument, taking into account that the linear equations $(2.78)_{\alpha}$, for $|\alpha| \geq 1$, are smoothly invertible in a fixed neighborhood of z_0.

q.e.d.

Lemmas 2.7 and 2.8 provide the necessary tools to reduce the system (2.70) $P = t\,\partial_t I_N - A(x,D_t,D_x)$ to some canonical form near $\dot{N}^*X$.

Fix a point $\rho_0 = (0,x_0, \pm 1,0) \in \dot{N}^*X$; without loss of generality we can suppose that the matrix $a_o(\rho_o)$, a_o being the principal symbol of A, is block diagonal: $a_o(\rho_o) = (a_{jj}(\rho_o))_{j=1,\dots,k}$, $k \geq 1$, with blocks $a_{jj}(\rho_o)$ of dimension $N_j \times N_j$, $N_1+\dots+N_k = N$, in such a way that if $k \geq 2$ then:

$$\sigma(a_{jj}(\rho_o) + r\, I_{N_j}) \cap \sigma(a_{\ell\ell}(\rho_o)) = \emptyset , \quad \begin{cases} j,\ell = 1,\dots,k,\ j \neq \ell, \\ r \in \mathbb{Z}\ . \end{cases} \tag{2.81}$$

When $k \geq 2$ we can apply the following result.

PROPOSITION 2.3. Let $\rho_o \in \dot{N}^*X$ and suppose that $a_o(\rho_o)$ has the above block structure with $k \geq 2$.

Then there exist two pdo's $E, \tilde{A} \in L^o(\tilde{X};\ N \times N)$ such that:

i) $E = E(x,D_t,D_x)$ is independent of t and elliptic near ρ_o.

ii) $\tilde{A} = \tilde{A}(x,D_t,D_x)$ is independent of t and is block diagonal $\tilde{A}(x,D_t,D_x) = (\tilde{A}_{jj}(x,D_t,D_x))_{j=1,\dots,k}$, with blocks of dimension $N_j \times N_j$, $j=1,\dots,k$. Moreover, $\tilde{a}^{(o)}_{jj}(\rho_o) = a_{jj}(\rho_o)$, $j=1,\dots,k$, where $\tilde{a}^{(o)}_{jj}$ denotes the principal symbol of $\tilde{A}_{jj}$.

iii)

$$(t\,\partial_t I_N - A(x,D_t,D_x))E \equiv E(t\,\partial_t I_N - \tilde{A}(x,D_t,D_x)) , \tag{2.82}$$

in a conic neighborhood of ρ_o.

Proof. We look for $E \sim E_o + E_{-1} + \dots$ and $\tilde{A} \sim \tilde{A}_o + \tilde{A}_{-1} + \dots$, with $E_{-j}(x,D_t,D_x)$, $\tilde{A}_{-j}(x,D_t,D_x) \in L^{-j}(\tilde{X};\ N \times N)$ for $j \geq 0$, such that conditions i) - iii) are satisfied.

To determine $E_o, \tilde{A}_o$ we require that their principal symbols, $e_o, \tilde{a}_o$ satisfy:

$$(2.83) \qquad -\tau \frac{\partial}{\partial\tau} e_o(x,\tau,\xi) - \{a_o(x,\tau,\xi)e_o(x,\tau,\xi) - e_o(x,\tau,\xi)\tilde{a}_o(x,\tau,\xi)\} = 0$$

in a conic neighborhood of ρ_o.

Suppose that $\rho_o = (0,x_o,1,0) \in N_+^*X$ (the case $\rho_o \in N_-^*X$ is treated in the same way). Near ρ_o we can write

$$\begin{cases} e_o(x,\tau,\xi) = e_o(x,1,\xi/\tau) = \hat{e}_o(x,\xi/\tau), \\ a_o(x,\tau,\xi) = a_o(x,1,\xi/\tau) = \hat{a}_o(x,\xi/\tau), \\ \tilde{a}_o(x,\tau,\xi) = \tilde{a}_o(x,1,\xi/\tau) = \hat{\tilde{a}}_o(x,\xi/\tau). \end{cases}$$

For $\hat{e}_o$ and $\hat{\tilde{a}}_o$ equation (2.83) becomes:

$$(2.84) \qquad (\eta \cdot \nabla_\eta \hat{e}_o)(x,\xi/\tau) - \{\hat{a}_o(x,\xi/\tau)\hat{e}_o(x,\xi/\tau) - \hat{e}_o(x,\xi/\tau)\hat{\tilde{a}}_o(x,\xi/\tau)\} = 0.$$

Taking into account (2.81) we can apply Lemma 2.7 with $M = X$, and find two smooth $N \times N$ matrices $\hat{e}_o(x,\eta)$, $\hat{\tilde{a}}_o(x,\eta)$ defined on $X \times \mathbb{R}^n$ such that:

i) $\hat{e}_o(x_o,0) = I_N$;

ii) $\hat{\tilde{a}}_o(x,\eta)$ is block diagonal with k blocks $\hat{\tilde{a}}_{jj}(x,\eta)$, $j = 1,\dots,k$ of dimension $N_j \times N_j$, for which $\hat{\tilde{a}}_o(x_o,0) = a_o(x_o,1,0)$;

iii) $(\eta \cdot \nabla_\eta \hat{e}_o)(x,\eta) - \{\hat{a}_o(x,\eta)\hat{e}_o(x,\eta) - \hat{e}_o(x,\eta)\hat{\tilde{a}}_o(x,\eta)\} = 0$ on a neighborhood $U \times (B_\varepsilon = \{\eta \in \mathbb{R}^n, |\eta| < \varepsilon\})$ of $(x_o,0)$.

Define the symbols $e_o(x,\tau,\xi) = \hat{e}_o(x,\xi/\tau)$, $\tilde{a}_o(x,\tau,\xi) = \hat{\tilde{a}}_o(x,\xi/\tau)$ for $x \in X$, $\xi \in \mathbb{R}^n$, $\tau > 0$. It follows that equation (2.83) is satisfied in the open cone

$U \times \{(\tau,\xi) \in \mathbb{R}^{n+1} \mid \tau > 0,\ |\xi| < \varepsilon\tau\}$, which is a conic neighborhood of ρ_o in $\dot{T}^*X$. Let $B_r(x_o) = \{x \in X \mid |x - x_o| < r\}$ be contained in U and fix $r' \in (\frac{r}{2}, r)$, $\varepsilon' \in (\frac{\varepsilon}{2}, \varepsilon)$.

We can modify e_o and $\tilde{a}_o$ outside of $U \times \{(\tau,\xi) \mid \tau > 0,\ |\xi| < \varepsilon\tau\}$ to obtain symbols $e_o(x,\tau,\xi)$, $\tilde{a}_o(x,\tau,\xi)$ defined on $\dot{T}^*X$ and equal to the old ones on $B_{r'}(x_o) \times \{(\tau,\xi) \mid \tau > 0,\ |\xi| < \varepsilon'\tau\}$. Taking $E_o = e_o(x,D_t,D_x)$ and $\tilde{A}_o = \tilde{a}_o(x,D_t,D_x)$, we see that E_o is elliptic near ρ_o, $\tilde{A}_o$ has the required matrix structure and $(t\,\partial_t I_N - A(x,D_t,D_x))E_o - E_o(t\,\partial_t I - \tilde{A}_o(x,D_t,D_x))$ belongs to $L^{-1}(\tilde{X};\ N \times N)$ on the open cone $x \in B_{r'}(x_o)$, $\tau > 0$, $|\xi| < \varepsilon'\tau$.

Suppose we have already constructed $E_{-j}(x,D_t,D_x)$, $\tilde{A}_{-j}(x,D_t,D_x)$ in $L^{-j}(\tilde{X};\ N \times N)$ for $j = 1,\dots,h$, with $\tilde{A}_{-j}$ block diagonal as required, such that $(t\,\partial_t I_N - A(x,D_t,D_x))(E_o+\dots+E_{-h}) - (E_o+\dots+E_{-h})(t\,\partial_t I_N - (\tilde{A}_o+\dots+\tilde{A}_{-h}))$ belongs to $L^{-(h+1)}(\tilde{X};\ N \times N)$ on some open cone $B_{r''}(x_o) \times \{(\tau,\xi) \mid \tau > 0,\ |\xi| < \varepsilon''\tau\}$ with $r'' \in (\frac{r}{2}, r)$, $\varepsilon'' \in (\frac{\varepsilon}{2}, \varepsilon)$.

We want to construct $E_{-(h+1)}(x,D_t,D_x)$ and $\tilde{A}_{-(h+1)}(x,D_t,D_x)$ in $L^{-(h+1)}(\tilde{X};\ N \times N)$ with $\tilde{A}_{-(h+1)}$ having the prescribed block diagonal structure, and such that

$$(t\,\partial_t I_N - A)\left\{\sum_{o}^{h+1}{}_j E_{-j}\right\} - \left\{\sum_{o}^{h+1}{}_j E_{-j}\right\}(t\,\partial_t I_N - \sum_{o\,j}^{h+1}\tilde{A}_{-j})$$

belongs to $L^{-(h+2)}(\tilde{X};\ N \times N)$ possibly in a narrower open cone

$$B_{\bar{r}}(x_o) \times \{(\tau,\xi) \mid \tau > 0,\ |\xi| < \bar{\varepsilon}\tau\} \text{ for some } \bar{r} \in (\tfrac{r}{2}, r''),\ \bar{\varepsilon} \in (\tfrac{\varepsilon}{2}, \varepsilon'').$$

We are led to the following equation for the principal symbols $e_{-(h+1)}$, $\tilde{a}_{-(h+1)}$

$$- \tau\, \partial_\tau\, e_{-(h+1)}(x,\tau,\xi) - \{a_o(x,\tau,\xi)\, e_{-(h+1)}(x,\tau,\xi)$$

$$(2.85)\qquad - e_{-(h+1)}(x,\tau,\xi)\, \tilde{a}_o(x,\tau,\xi)\} + e_o(x,\tau,\xi)\, \tilde{a}_{-(h+1)}(x,\tau,\xi) =$$

$$= \varphi_{-(h+1)}(x,\tau,\xi) \ ,$$

where $\varphi_{-(h+1)}(x,\tau,\xi)$ is positively homogeneous of degree $-(h+1)$ in the open cone $B_{r''}(x_o) \times \{(\tau,\xi) \mid \tau > 0, \quad |\xi| < \varepsilon''\tau\}$.

Again, we observe that for $\tau > 0$:

$$\begin{cases} e_{-(h+1)}(x,\tau,\xi) = \tau^{-(h+1)} e_{-(h+1)}(x,1,\xi/\tau) = \tau^{-(h+1)} \hat{e}_{-(h+1)}(x,\xi/\tau) \\ a_o(x,\tau,\xi) = a_o(x,1,\xi/\tau) = \hat{a}_o(x,\xi/\tau) \\ \tilde{a}_o(x,\tau,\xi) = \tilde{a}_o(x,1,\xi/\tau) = \tilde{\hat{a}}_o(x,\xi/\tau) \\ e_o(x,\tau,\xi) = e_o(x,1,\xi/\tau) = \hat{e}_o(x,\xi/\tau) \\ \tilde{a}_{-(h+1)}(x,\tau,\xi) = \tau^{-(h+1)} \tilde{a}_{-(h+1)}(x,1,\xi/\tau) = \tau^{-(h+1)} \tilde{\hat{a}}_{-(h+1)}(x,\xi/\tau) \\ \varphi_{-(h+1)}(x,\tau,\xi) = \tau^{-(h+1)} \varphi_{-(h+1)}(x,1,\xi/\tau) = \tau^{-(h+1)} \hat{\varphi}_{-(h+1)}(x,\xi/\tau) \end{cases}$$

Equation (2.85) can be rewritten as :

$$[(h+1) I_N + \eta \cdot \nabla_\eta]\, \hat{e}_{-(h+1)}(x,\eta) - \{\hat{a}_o(x,\eta)\, \hat{e}_{-(h+1)}(x,\eta)$$

$$(2.86)\qquad - \hat{e}_{-(h+1)}(x,\eta)\, \tilde{\hat{a}}_o(x,\eta)\} + \hat{e}_o(x,\eta)\, \tilde{\hat{a}}_{-(h+1)}(x,\eta) = \hat{\varphi}_{-(h+1)}(x,\eta) \ ,$$

with $\eta = \xi/\tau$.

Application of Lemma 2.7 (part 2-) yields the existence of smooth $N \times N$ matrices $\hat{e}_{-(h+1)}(x,\eta)$, $\tilde{\hat{a}}_{-(h+1)}(x,\eta)$ defined on $X \times \mathbb{R}^n$ such that equation (2.86) is satisfied on $U \times B_\varepsilon$.

Putting $e_{-(h+1)}(x,\tau,\xi) = \tau^{-h-1}\hat{e}_{-(h+1)}(x,\xi/\tau)$ and $\tilde{a}_{-(h+1)}(x,\tau,\xi) =$

$= \tau^{-h-1}\tilde{\hat{a}}_{-(h+1)}(x,\xi/\tau)$ we obtain smooth symbols defined for $\tau > 0$, with $\tilde{a}_{-(h+1)}(x,\tau,\xi)$ having the correct block diagonal structure, such that equation (2.85) is satisfied on the open cone $B_{r'''}(x_o) \times \{(\tau,\xi) \mid \tau > 0,\ |\xi| < \varepsilon''' \tau\}$ for some $r''' \in (\frac{r}{2}, r'')$, $\varepsilon''' \in (\frac{\varepsilon}{2}, \varepsilon'')$.

We modify $e_{-(h+1)}$, $\tilde{a}_{-(h+1)}$ to obtain symbols defined on $\dot{T}^*X$ and equal to the old ones on some open cone around ρ_o still containing

$\bar{B}_{r/2}(x_o) \times \{(\tau,\xi) \mid \tau > 0,\ |\xi| \leqq \varepsilon/2\ \tau\}$.

Using induction on h and defining $E \sim E_o + E_{-1} + \ldots$, $\tilde{A} \sim \tilde{A}_o + \tilde{A}_{-1} + \ldots$, we are finished.

q.e.d.

The preceding result reduces the study of system (2.70) near N^*_+X (or N^*_-X) to the study of a finite number of decoupled systems $t\partial_t I_{N_i} - \tilde{A}_{ii}(x,D_t,D_x)$, $i = 1,\ldots,k$, such that, denoting by $\tilde{a}^{(o)}_{ii}(x,\tau,\xi)$ the principal symbol of $\tilde{A}_{ii}$, we have at a point $\rho_o \in N^*_+X$ (or $\rho_o \in N^*_-X$):

1) $\sigma(\tilde{a}^{(o)}_{ii}(\rho_o) + r\, I_{N_i}) \cap \sigma(\tilde{a}^{(o)}_{jj}(\rho_o)) = \emptyset$,

for $i,j = 1,\ldots,k$, $i \neq j$, and for every $r \in \mathbb{Z}$.

2) $\tilde{a}^{(o)}_{ii}(\rho_o)$ is in Jordan canonical form. Precisely, either

$$(2.87)\quad \tilde{a}_{ii}^{(o)}(\rho_o) = \begin{pmatrix} \begin{matrix} \lambda_i & & * \\ & \ddots & \\ 0 & & \lambda_i \end{matrix} & & & 0 \\ & \begin{matrix} \lambda_i - \ell_{1i} & & * \\ & \ddots & \\ 0 & & \lambda_i - \ell_{1i} \end{matrix} & & \\ & & \ddots & \\ 0 & & & \begin{matrix} \lambda_i - \ell_{h_i i} & & * \\ & \ddots & \\ 0 & & \lambda_i - \ell_{h_i i} \end{matrix} \end{pmatrix}, \quad i = 1,\dots,k \,,$$

where $\lambda_i \in \mathbb{C}$ and $\ell_{ji} \in \mathbb{N}$, $j = 1,\dots,h_i$ with $1 \leqq \ell_{1i} < \ell_{2i} < \dots < \ell_{h_i i}$, $h_i \geqq 1$; or:

$$(2.88)\quad \tilde{a}_{ii}^{(o)}(\rho_o) = \begin{pmatrix} \lambda_i & & * \\ & \ddots & \\ 0 & & \lambda_i \end{pmatrix}$$

where $\lambda_i \in \mathbb{C}$.

We shall consider the systems $t\,\partial_t I_{N_i} - \tilde{A}_{ii}(x,D_t,D_x)$ more closely. To symplify the notation we drop the indices i and consider a system :

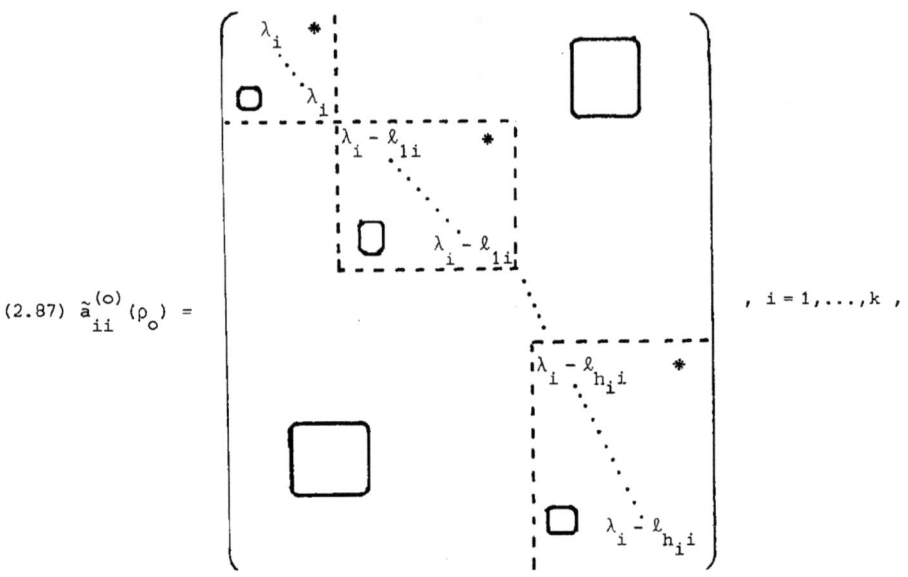

$$(t\,\partial_t I_N - A(x,D_t,D_x))u = f\ , \tag{2.89}$$

where, at some point $\rho_o \in \dot{N}^*X$, the principal symbol $a_o(x,\tau,\xi)$ of the operator $A \in L^o(\tilde{X};\ N \times N)$ is written in Jordan canonical form of one of the following types :

either

(2.90) $a_o(\rho_o) =$

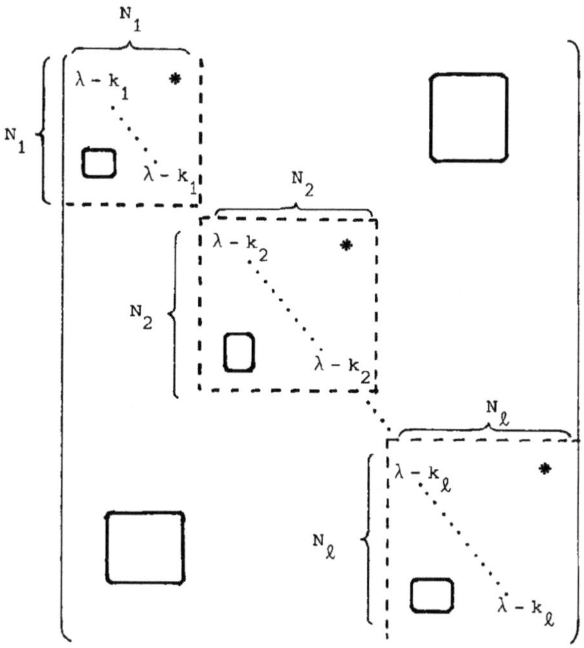

where $\lambda \in \mathbb{C}$ and $0 = k_1 < k_2 <\ldots< k_\ell$ are some non-negative integers, $\ell \geq 2$, $N_1+\ldots+N_\ell = N$, or

$$(2.91) \quad a_o(\rho_o) = \begin{pmatrix} \lambda & & & * \\ & \lambda & & \\ & & \ddots & \\ \square & & & \lambda \end{pmatrix} ,$$

where $\lambda \in \mathbb{C}$.

We have the following result.

PROPOSITION 2.4. Consider the system (2.89) where $a_o(\rho_o)$ has the form (2.90) for some $\rho_o \in \dot{N}^*X$.

Then there exist:

1) A pdo $E = E(x,D_t,D_x) \in L^o(\tilde{X};\ N \times N)$ elliptic near ρ_o;

2) A pdo $C = C(x,D_t,D_x) \in L^o(\tilde{X};\ N \times N)$ with the following matrix structure:

$$C(x,D_t,D_x) = (C_{ij}(x,D_t,D_x))_{i,j=1,\dots,\ell}$$

with blocks $C_{ij}(x,D_t,D_x)) \in L^o(\tilde{X};\ N_i \times N_j), i,j = 1,\dots,\ell$, and

α) For $i > j$, $C_{ij}(x,D_t,D_x) = 0$.

β) For $i = j$, $C_{jj}(x,D_t,D_x) = C_{jj}^{(o)}(x)$ i.e. the full symbol of C_{jj} reduces to a smooth matrix $C_{jj}^{(o)}(x)$ depending only on x, $j = 1,\dots,\ell$.

γ) For $i < j$, the full symbol of $C_{ij}(x,D_t,D_x)$ is of the form

$$\sum_{o}^{k_j-k_i}{}_h \sum_{|\alpha| = k_j - k_i - h} C^{(\alpha)}_{ij,-h}(x) \frac{1}{\tau^h} \left(\frac{\xi}{\tau}\right)^\alpha$$ in a conic neighborhood of ρ_o,

for some polynomial $\sum_{o}^{k_j - k_i} {}_h \sum_{|\alpha| = k_j - k_i - h} C^{(\alpha)}_{ij,-h}(x)\eta^{\alpha}$ whose coefficients are smooth $N_i \times N_j$ matrices depending only on x.

δ) If $\rho_o = (0,x_o, \pm 1, 0)$, then

$$(C^{(o)}_{jj}(x_o))_{j=1,\dots,\ell} = a_o(\rho_o) ,$$

such that

(2.92) $$(t\,\partial_t I_N - A(x,D_t,D_x))E \equiv E(t\,\partial_t I_N - C(x,D_t,D_x), \quad \text{near} \quad \rho_o .$$

When $a_o(\rho_o)$ is of the form (2.91), we have (2.92) for some $E = E(x,D_t,D_x) \in L^{o}(\tilde{X};\ N \times N)$ elliptic near ρ_o with a $C(x,D_t,D_x) \in L^{o}(\tilde{X};\ N \times N)$ whose full symbol reduces to a smooth $N \times N$ matrix $C(x)$ depending only on x and satisfying $C(x_o) = a_o(\rho_o)$.

Proof. Assume that $\rho_o = (0,x_o,1,0) \in N^*_+X$ (the case $\rho_o \in N^*_-X$ can be treated similarly). Suppose that $a_o(\rho_o)$ has the form (2.90). We look for operators $E \sim E_o + E_{-1} + \dots$ and $C \sim C_o + C_{-1} + \dots$, with $E_{-h}(x,D_t,D_x)$ and $C_{-h}(x,D_t,D_x) \in L^{-h}(\tilde{X};\ N \times N)$ for $h \geq 0$, such that:

i) E_o is elliptic near ρ_o;

ii) $C_{-h}(x,D_t,D_x) = (C_{ij,-h}(x,D_t,D_x))_{i,j=1,\dots,\ell}$, with blocks $C_{ij,-h} \in L^{-h}(\tilde{X};\ N_i \times N_j)$ satisfying:

a) For $i > j$ and for every $h \geq 0$, $C_{ij,-h}(x,D_t,D_x) = 0$.

b) For $i \leq j$ and $h = 0$, the full symbol of the operator $C_{ij,0}(x,D_t,D_x)$ is of the form $\sum_{|\alpha| = k_j - k_i} c^{(\alpha)}_{ij,0}(x) \left(\frac{\xi}{\tau}\right)^{\alpha}$ in a conic neighborhood of ρ_o, for some polynomial $\sum_{|\alpha| = k_j - k_i} c^{(\alpha)}_{ij,0}(x)\eta^{\alpha}$ whose coefficients are smooth $N_i \times N_j$ matrices, for which $(c^{(o)}_{ij,0}(x_o))_{j=1,\dots,\ell} = a_o(\rho_o)$. For $h = 1,2,\dots$ and for $i \leq j$, the full symbol of the operator $C_{ij,-h}(x,D_t,D_x)$ in a conic neighborhood of ρ_o is of the form:

$$c_{ij,-h}(x,\tau,\xi) = \begin{cases} 0 & , \text{ if } k_j - k_i - h < 0 , \\ \sum_{|\alpha| = k_j - k_i - h} c^{(\alpha)}_{ij,-h}(x) \frac{1}{\tau^h} \left(\frac{\xi}{\tau}\right)^{\alpha} , & \text{if } k_j - k_i - h \geq 0 , \end{cases}$$

for some polynomial $\sum_{|\alpha| = k_j - k_i - h} c^{(\alpha)}_{ij,-h}(x)\eta^{\alpha}$ whose coefficients are smooth $N_i \times N_j$ matrices.

iii) $(t\,\partial_t I_N - A(x,D_t,D_x))E \equiv E(t\,\partial_t I_N - C(x,D_t,D_x))$, near ρ_o.

First we try to construct E_o and C_o so that $(t\,\partial_t I_N - A)E_o - E_o(t\,\partial_t I_N - C_o) \in L^{-1}(\tilde{X};\, N \times N)$ in some open cone around ρ_o. We are led to the equation:

$$(2.93) \qquad -\tau\,\partial_\tau\, e_o(x,\tau,\xi) - \{a_o(x,\tau,\xi)e_o(x,\tau,\xi) - e_o(x,\tau,\xi)c_o(x,\tau,\xi)\} = 0 .$$

As in Proposition 2.3, write, near ρ_o ,

$$\begin{cases} e_o(x,\tau,\xi) = e_o(x,1,\xi/\tau) = \hat{e}_o(x,\xi/\tau), \\ a_o(x,\tau,\xi) = a_o(x,1,\xi/\tau) = \hat{a}_o(x,\xi/\tau), \\ c_o(x,\tau,\xi) = c_o(x,1,\xi/\tau) = \hat{c}_o(x,\xi/\tau). \end{cases}$$

Equation (2.93) becomes:

(2.94) $$(\eta \cdot \nabla_\eta \hat{e}_o)(x,\xi/\tau) - \{\hat{a}_o(x,\xi/\tau)\hat{e}_o(x,\xi/\tau) - \hat{e}_o(x,\xi/\tau)\hat{c}_o(x,\xi/\tau)\} = 0 .$$

Using Lemma 2.8, 1- with $M = X$ we can find two smooth $N \times N$ matrices $\hat{e}_o(x,\eta)$, $\hat{c}_o(x,\eta)$ defined on $X \times \mathbb{R}^n_\eta$ such that:

i) $\hat{e}_o(x_o,0) = I_N$;

ii) $\hat{c}_o(x,\eta) = (\hat{c}_{ij}(x,\eta))_{i,j=1,\dots,\ell}$, with blocks $\hat{c}_{ij}$ of dimension $N_i \times N_j$ for which:

1) $\hat{c}_{ij} = 0$, if $i > j$.

2) $\hat{c}_{jj}(x,\eta) = c^{(o)}_{jj}(x)$ is independent of η, $j = 1,\dots,\ell$, and $(c^{(o)}_{jj}(x_o))_{j=1,\dots,\ell} = a_o(\rho_o)$.

3) $\hat{c}_{ij}(x,\eta) = \sum_{|\alpha| = k_j - k_i} c^{(\alpha)}_{ij,0}(x)\eta^\alpha$, if $i < j$, for some smooth $N_i \times N_j$ matrices $c^{(\alpha)}_{ij,0}(x)$.

iii) equation (2.94) is satisfied on a neighborhood $U \times (B_\varepsilon = \{\eta \mid |\eta| < \varepsilon\})$ of $(x_0, 0)$.

Define the symbols $e_o(x,\tau,\xi) = \hat{e}_o(x, \frac{\xi}{\tau})$, $c_o(x,\tau,\xi) = \hat{c}_o(x, \frac{\xi}{\tau})$ for $x \in X$, $\xi \in \mathbb{R}^n$, $\tau > 0$. It follows that equation (2.93) is satisfied on the cone $U \times (\Gamma_\varepsilon = \{(\tau,\xi) \in \mathbb{R}^{n+1} \mid \tau > 0, |\xi| < \varepsilon\tau\})$. Let $B_r(x_0) = \{x \in X \mid |x-x_0| < r\} \subset U$ and fix $r' \in (\frac{r}{2}, r)$, $\varepsilon' \in (\varepsilon/2, \varepsilon)$. We modify e_o and c_o outside $U \times \Gamma_\varepsilon$ to obtain smooth symbols on $\dot{T}^*\tilde{X}$ equal to the old ones on $B_{r'}(x_0) \times \Gamma_{\varepsilon'}$.
Then we define $E_o = e_o(x,D_t,D_x)$, $C_o = c_o(x,D_t,D_x)$. We see that E_o is elliptic near ρ_o, C_o has the required matrix structure and $(t\,\partial_t I_N - A)E_o - E_o(t\,\partial_t I_N - C_o) \in L^{-1}$ on the open cone $B_{r'}(x_0) \times \Gamma_{\varepsilon'}$.
Suppose we have already constructed $E_o, E_{-1}, \ldots, E_{-(h-1)}$ and $C_o, C_{-1}, \ldots, C_{-(h-1)}$ with the correct matrix structure such that

$$(t\,\partial_t I_N - A)\left(\sum_{o}^{h-1}{}_j E_{-j}\right) - \left(\sum_{o}^{h-1}{}_j E_{-j}\right)\left(t\,\partial_t I_N - \sum_{o}^{h-1}{}_j C_{-j}\right) \in L^{-h}(\tilde{X};\, N \times N)$$

on some open cone $B_{\bar{r}}(x_0) \times \Gamma_{\bar{\varepsilon}}$, with $\bar{r} \in (\frac{r}{2}, r)$, $\bar{\varepsilon} \in (\frac{\varepsilon}{2}, \varepsilon)$.
To find E_{-h} and C_{-h} we consider the corresponding transport equation:

$$\begin{aligned} &-\tau\,\partial_\tau e_{-h}(x,\tau,\xi) - \{a_o(x,\tau,\xi)e_{-h}(x,\tau,\xi) - e_{-h}(x,\tau,\xi)c_o(x,\tau,\xi)\} \\ &+ e_o(x,\tau,\xi)c_{-h}(x,\tau,\xi) = g_{-h}(x,\tau,\xi), \end{aligned} \tag{2.95}$$

where $g_{-h}(x,\tau,\xi)$ is a smooth function positively homogeneous of degree $-h$ in the open cone $B_{\bar{r}}(x_0) \times \Gamma_{\bar{\varepsilon}}$. For $\tau > 0$ we have :

$$\begin{cases} e_{-h}(x,\tau,\xi) = \tau^{-h}\, e_{-h}(x,1,\xi/\tau) = \tau^{-h}\, \hat{e}_{-h}(x,\xi/\tau) \\ c_{-h}(x,\tau,\xi) = \tau^{-h}\, c_{-h}(x,1,\xi/\tau) = \tau^{-h}\, \hat{c}_{-h}(x,\xi/\tau) \\ g_{-h}(x,\tau,\xi) = \tau^{-h}\, g_{-h}(x,1,\xi/\tau) = \tau^{-h}\, \hat{g}_{-h}(x,\xi/\tau) \\ a_o(x,\tau,\xi) = a_o(x,1,\xi/\tau) = \hat{a}_o(x,\xi/\tau) \\ e_o(x,\tau,\xi) = e_o(x,1,\xi/\tau) = \hat{e}_o(x,\xi/\tau) \\ c_o(x,\tau,\xi) = c_o(x,1,\xi/\tau) = \hat{c}_o(x,\xi/\tau). \end{cases}$$

Equation (2.95) can be rewritten as :

$$\text{(2.96)}\quad \begin{aligned} &(h\, I_N + \eta \cdot \nabla_\eta)(\hat{e}_{-h})(x,\eta) \\ &- \{\hat{a}_o(x,\eta)\hat{e}_{-h}(x,\eta) - \hat{e}_{-h}(x,\eta)\hat{c}_o(x,\eta)\} + \hat{e}_o(x,\eta)\hat{c}_{-h}(x,\eta) = \hat{g}_{-h}(x,\eta) \ , \end{aligned}$$

with $\eta = \xi/\tau$.

Application of Lemma 2.8, 2 - yields the existence of smooth $N \times N$ matrices $\hat{e}_{-h}(x,\eta)$, $\hat{c}_{-h}(x,\eta)$ defined on $X \times \mathbb{R}^n$, such that $\hat{c}_{-h}(x,\eta) = (\hat{c}_{ij,-h}(x,\eta))_{i,j=1,\ldots,}$ with blocks $\hat{c}_{ij,-h}$ of dimension $N_i \times N_j$ for which

1) $\hat{c}_{ij,-h} = 0$ if $i > j$.

2) if $i \leqq j$ and $k_j - k_i - h < 0$, then $\hat{c}_{ij,-h} = 0$; if $i \leqq j$ and $k_j - k_i - h \geqq 0$, then $\hat{c}_{ij,-h}(x,\eta) = \sum_{|\alpha| = k_j - k_i - h} c^{(\alpha)}_{ij,-h}(x)\eta^\alpha$ for some smooth $N_i \times N_j$ matrices $c^{(\alpha)}_{ij,-h}(x)$.

Moreover, equation (2.96) is satisfied on $U \times B_\varepsilon$. Putting $e_{-h}(x,\tau,\xi) = \tau^{-h}\, \hat{e}_{-h}(x,\xi/\tau)$ and $c_{-h}(x,\tau,\xi) = \tau^{-h}\, \hat{c}_{-h}(x,\xi/\tau)$, we obtain smooth symbols defined for $\tau > 0$, with c_{-h} having the correct matrix structure, such that equation (2.93) is satisfied on the open cone $B_{r''}(x_o) \times \Gamma_{\varepsilon''}$, for some $r'' \in (\frac{r}{2}, \bar{r})$, $\varepsilon'' \in (\frac{\varepsilon}{2}, \bar{\varepsilon})$.

Modifying e_{-h}, c_{-h} we obtain smooth symbols defined on $\dot{T}^*\tilde{X}$ and equal to the old ones on some open cone around ρ_o containing $\bar{B}_{r/2}(x_o) \times \overline{\Gamma_{\varepsilon/2}}$. Using induction on h and defining $E \sim \sum_{j \geq 0} E_{-j}$, $C \sim \sum_{j \geq 0} C_{-j}$, we are finished.

The proof in the case $a_o(\rho_o)$ has the form (2.91) goes in the same way as above and we leave the details to the reader.

q.e.d.

In the next result we show that a system of the form $t\,\partial_t I_N - C(x)$, $C(x)$ being a smooth $N \times N$ matrix defined on X, can be reduced microlocally near N^*_+X (or N^*_-X) to the form $t\,I_N$. To do this we need some notation. For every $\delta \in [0,1)$ we denote by $L^m_{1,\delta}(\tilde{X};\, N \times N)$ the set of all $N \times N$ matrices of properly supported p d o's with symbols in $S^m_{1,\delta}(\tilde{X})$ (Cfr. Hörmander [14]); we recall that $WF(Qu) \subset WF(u)$ for every $Q \in L^m_{1,\delta}(\tilde{X};\, N \times N)$ and for all $u \in \mathcal{D}'(\tilde{X})^N$.

Let $G(x)$ be a smooth $N \times N$ matrix defined on X and let m be any number $> \sup_{x \in X}$ (spectral radius $G(x)$). Denote by $D^{\pm iG(x)}_{t,+} \in L^m_{1,\delta}(\tilde{X};\, N \times N)$, $0 < \delta < 1$, the operator with full symbol $= \tau^{\pm iG(x)}$ in a conic neighborhood of N^*_+X and rapidly decreasing in a conic neighborhood of N^*_-X. In the same way, denote by $D^{\pm iG(x)}_{t,-} \in L^m_{1,\delta}(\tilde{X};\, N \times N)$, $0 < \delta < 1$, the operator with full symbol $= \tau^{\pm iG(x)}$ in a conic neighborhood of N^*_-X and rapidly decreasing in a conic neighborhood of N^*_+X.

Note that:

$$(2.97)\quad \begin{cases} D_{t,+}^{\pm iG(x)} \; D_{t,+}^{\mp iG(x)} \equiv I_N , & \text{near } N^*_+X , \\ D_{t,-}^{\pm iG(x)} \; D_{t,-}^{\mp iG(x)} \equiv I_N , & \text{near } N^*_-X . \end{cases}$$

PROPOSITION 2.5. Let the system $Pu = (t\,\partial_t I_N - G(x))u = f$ be given, where $G(x)$ is a smooth $N \times N$ matrix defined on X and satisfying $\sup_{x \in X} \| G(x) \| = m < \infty$.

Then :

$$(2.98)\quad \begin{cases} D_{t,+}^{iG(x)} (t\,\partial_t I_N - G(x)) \frac{1}{i} D_{t,+}^{-iG(x)-I_N} \equiv t\, I_N , & \text{near } N^*_+X , \\ D_{t,-}^{iG(x)} (t\,\partial_t I_N - G(x)) \frac{1}{i} D_{t,-}^{-iG(x)-I_N} \equiv t\, I_N , & \text{near } N^*_-X . \end{cases}$$

As a consequence, for every $\rho_0 = (0, x_0, \pm 1, 0) \in N^*X$, denoting by $P_{\rho_0} : \mathcal{M}_{\rho_0}(\tilde{X})^N \longrightarrow \mathcal{M}_{\rho_0}(\tilde{X})^N$ the linear map induced by P, one has:

$$(2.99)\qquad \ker P_{\rho_0} \simeq \mathcal{D}'_{x_0}(X)^N , \; \text{Coker } P_{\rho_0} = \{0\} ,$$

where $\mathcal{D}'_{x_0}(X)$ is the space of germs of distributions in the x-variables defined in a neighborhood of x_0 .

<u>Proof</u>. (2.98) follows by a direct computation. To prove the second part it is enough to show that for the map $t\, I_N : \mathcal{M}_{\rho_0}(\tilde{X})^N \longrightarrow \mathcal{M}_{\rho_0}(\tilde{X})^N$ we have $\ker \; t\, I_N \simeq \mathcal{D}'_{x_0}(X)^N$, Coker $t\, I_N = \{0\}$.

Denote by $\pi_{\rho_0} : M(\tilde{X})^N \longrightarrow M_{\rho_0}(\tilde{X})^N$ the canonical surjection; we can suppose that $N = 1$ and $\rho_0 \in \dot{N}^*_+X$. Given $\hat{f} \in M_{\rho_0}(\tilde{X})$ there exists a distribution $f \in D'(\tilde{X})$ such that $\pi_{\rho_0}(f) = \hat{f}$ and $WF(f)$ is concentrated in a conic neighborhood of ρ_0 disjoint from N^*_-X. Since the product $\left(\frac{1}{t+i0} \otimes 1_x\right) f(t,x)$ is well defined, taking $\hat{u} = \pi_{\rho_0}\left(\left(\frac{1}{t+i0} \otimes 1_x\right) f\right)$, we have $t\hat{u} = \hat{f}$.

To prove that $\ker t \simeq D'_{x_0}(X)$, suppose that $\hat{u} \in M_{\rho_0}(\tilde{X})$ and $t\hat{u} = \hat{0}$. Consider a distribution $u \in D'(\tilde{X})$ for which $WF(u) \cap N^*_-X = \emptyset$ and $\pi_{\rho_0}(u) = \hat{u}$. Since $(0,x_0,\pm 1,0) \notin WF(tu)$, there exists a distribution $\tilde{u}(x)$ defined on a neighborhood U of x_0 such that $tu|_{t=0} = \tilde{u}$ on U.

Denote by $[\tilde{u}]$ the germ of $\tilde{u}$, $[\tilde{u}] \in D'_{x_0}(X)$. We observe that if $v \in D'(\tilde{X})$ and $\pi_\rho(v) = \hat{u}$, then $[\tilde{v}] = [\tilde{u}]$, since $v - u$ has a trace at $t = 0$ for x near x_0. We thus have a well defined linear map from $\ker t$ into $D'_{x_0}(X)$ which is clearly injective. If $\varphi(x)$ is a distribution defined near x_0 consider $u(t,x) = \frac{1}{t+i0} \otimes \varphi(x)$; then $t\,\pi_{\rho_0}(u) = \hat{0}$ and $[\tilde{u}] = [\varphi]$, so that the map $\ker t \longrightarrow D'_{x_0}(X)$ is surjective.

q.e.d.

We can now prove the main result.

THEOREM 2.3. Consider the system $P = t\partial_t I_N - A(t,x,D_t,D_x)$, with $A \in L^o(\tilde{X}; N \times N)$. For every $\rho_0 \in \dot{N}^*X$ denote by $P_{\rho_0} : M_{\rho_0}(\tilde{X})^N \longrightarrow M_{\rho_0}(\tilde{X})^N$ the linear map induced by P on the stalk over ρ_0 of the sheaf $M(\tilde{X})^N$ of (vector-valued) microdistributions on $\tilde{X}$.

Then $\ker P_{\rho_0} \simeq D'_{x_0}(X)^N$, $\operatorname{Coker} P_{\rho_o} = \{0\}$.

Proof. Using Lemma 2.6 and Proposition 2.3 we are reduced to prove the result for an operator $\tilde{P} = t\,\partial_t I_N - A(x,D_t,D_x)$ where the principal symbol $a_o(x,\tau,\xi)$ of the operator A has either the form (2.90) or the form (2.91) at ρ_o. When $a_o(\rho_o)$ has the form (2.91), we use Proposition 2.4 and find a p d o $E \in L^o(\tilde{X};\, N\times N)$, elliptic near ρ_o, such that $\tilde{P}\,E \equiv E(t\,\partial_t I_N - C(x))$, near ρ_o, where $C(x)$ is a smooth $N \times N$ matrix defined on X and satisfying $C(x_o) = a_o(\rho_o)$, if $\rho_o = (0,x_o,\pm 1,0)$. We can modify $C(x)$ outside of a neighborhood of x_o to obtain a matrix, still denoted by $C(x)$, satisfying $\sup_{x\in X} \| C(x) \| < \infty$. Application of Proposition 2.5 yields that $\ker P_{\rho_o} \simeq \mathcal{D}'_{x_o}(X)^N$ and $\mathrm{Coker}\, P_{\rho_o} = \{0\}$.

When $a_o(\rho_o)$ has the form (2.90), we use Proposition 2.4 and find two p d o's $E, C \in L^o(\tilde{X};\, N \times N)$ for which:

i) E is elliptic near ρ_o,

ii) $C = (C_{ij})_{i,j=1,\dots,\ell}$ with blocks of dimension $N_i \times N_j$, $N_1+\dots+N_\ell = N$, such that $C_{ij} = 0$ for $i > j$, C_{jj} reduces to a smooth matrix depending only on $x, j = 1,\dots,\ell$.

iii) $\tilde{P}\,E \equiv E(t\,\partial_t I - C)$, near ρ_o.

Put $\hat{P}_j = t\,\partial_t I_{N_j} - C_{jj}(x)$, $j = 1,\dots,\ell$. Since, by Proposition 2.5, $(\hat{P}_j)_{\rho_o} : M_{\rho_o}(\tilde{X})^{N_j} \longrightarrow M_{\rho_o}(\tilde{X})^{N_j}$ is surjective for every j, it is easily seen that

$$\ker P_{\rho_o} \simeq \bigoplus_{j=1}^{\ell} \ker(\hat{P}_j)_{\rho_o}\,. \tag{2.99}'$$

Applying Proposition 2.5 once more, we have $\ker(\hat{P}_j)_{\rho_o} \simeq \mathcal{D}'_{x_o}(X)^{N_j}$, $j = 1,\dots,\ell$, so that the conclusion follows. q.e.d.

Remarks. The structure of $\ker P_{\rho_0}$ can be made more precise if we suppose that the matrix $a_o(\rho_o)$ has N different eigenvalues $\lambda_j(\rho_o)$, $j=1,\dots,N$, such that $\lambda_i(\rho_o) - \lambda_j(\rho_o) \notin \mathbb{Z}$ for $i \neq j$. In this case, using Lemma 2.6 and Propositions 2.3, 2.4, we can find two p d o's $R_\pm, S_\pm \in L^o(\tilde{X}; N \times N)$, elliptic near $\rho_o \in N^*_\pm X$, such that :

$$(2.100) \qquad R_\pm(t\,\partial_t I_N - A(t,x,D_t,D_x))S_\pm \equiv (t\,\partial_t I_N - H_\pm(x)) , \quad \text{near } \rho_o ,$$

where

$$(2.101) \qquad H_\pm(x) = \begin{pmatrix} \lambda_1^\pm(x) & & \Box \\ & \ddots & \\ \Box & & \lambda_N^\pm(x) \end{pmatrix}$$

and the $\lambda_j^\pm(x)$ are C^∞ functions on X for $j=1,\dots,N$, which coincide with the N distinct eigenvalues of $a_o(0,x,\pm 1,0)$ in a neighborhood of x_o. From (2.100), (2.101) it follows that

$$(2.102) \qquad \ker P_{\rho_o} \simeq \bigoplus_{j=1}^{N} \ker (t\,\partial_t - \lambda_j^\pm(x)) .$$

Now, if $\lambda(x) \in C^\infty(X)$, it is easily recognized that

$$(2.103)\quad \rho_o \in N^*_{\pm}X \Rightarrow \ker(t\,\partial_t - \lambda(x))_{\rho_o} = \begin{cases} \{\pi_{\rho_o}((t \pm io)^{\lambda(x)} \otimes \varphi(x)) \mid \varphi \in \mathcal{D}'_{x_o}(X)\}, \\ \quad \text{if} \quad \lambda(x_o) \notin \mathbb{Z}_+ ; \\ \{\pi_{\rho}(t_{\pm}^{\lambda(x)} \otimes \varphi(x)) \mid \varphi \in \mathcal{D}'_{x_o}(X)\}, \\ \quad \text{if} \quad \lambda(x_o) \in \mathbb{Z}_+ . \end{cases}$$

From (2.103) and (2.101) one can derive the structure of $\ker P_{\rho_o}$. The above discussion as well as Theorem 2.3 should be compared with analogous results proved in the hyperfunction setting by H. Tahara [28, Theorem 2.2.14] and M. Kashiwara - - T. Oshima [19, Proposition 4.4] .

3. APPLICATIONS TO FUCHSIAN HYPERBOLIC P.D.E.

We are interested in studying operators of the form :

$$P = \sum_{j=0}^{k} t^{k-j} P_{m-j}(t,x;\, D_t, D_x) \;, \tag{3.1}$$

where k and m are positive integers $1 \leqq k \leqq m$, and $P_{m-j}(t,x,D_t,D_x)$ is a pdo defined in an open set $\tilde{X} = (-T,T) \times X \subset \mathbb{R}_t \times \mathbb{R}^n_x$, $j = 0,1,\ldots,k$. Denoting by $p_{m-j}(t,x;\tau,\xi)$ the principal symbol of P_{m-j}, we suppose that :

(i) $\quad p_m \big|_{\dot{N}^*X} \neq 0$;

(ii) There exists a real function $\lambda(t,x,\xi) \in C^\infty(\tilde{X} \times \mathbb{R}^n \setminus \{0\})$, positively homogeneous of degree 1 in ξ, such that the following factorization holds:

$$p_m(t,x,\tau,\xi) = (\tau - \lambda(t,x,\xi))^r e(t,x,\tau,\xi) \;, \tag{3.2}$$

where r is a positive integer, $e \in C^\infty(\tilde{X} \times \mathbb{R}^n \setminus \{0\})$ and $e(t,x,\lambda(t,x,\xi),\xi) \neq 0$ for every $(t,x,\xi) \in \tilde{X} \times \mathbb{R}^n \setminus \{0\}$.

(iii) When $r \geqq 2$, P satisfies the Levi condition :

$$
(L)\quad \begin{cases} \text{For each real smooth function } \varphi(t,x) \text{ satisfying} \\[4pt] \qquad \partial_t\varphi(t,x) - \lambda(t,x,d_x\varphi(t,x)) = 0 \\[4pt] \text{in a neighborhood of } (0,x_o) \in \tilde{X}\text{, with } d_x\,\varphi(0,x_o) \neq 0\text{, we have} \\[4pt] \qquad e^{-is\varphi(t,x)}P(a(t,x)e^{is\varphi(t,x)}) = O(s^{m-r}),\ s \to \infty\ , \\[4pt] \text{for every } a \in C_o^\infty(\tilde{X}) \text{ with } d_x\varphi \neq 0 \text{ on the support of } a\ . \end{cases}
$$

In this Section we will study the singularities of a solution $u \in \mathcal{D}'(\tilde{X})$ of the equation $Pu = f$ near a point $\rho_o = (0,x_o,\lambda(0,x_o,\xi^{(o)}),\ \xi^{(o)}) \in \dot{T}^*\tilde{X} \setminus N^*X$. Note that the principal symbol of P vanishes on the union of the two conic hypersurfaces:

$$
(3.3)\qquad \begin{cases} \Sigma_1 = \{(t,x,\tau,\xi) \in \dot{T}^*\tilde{X} \mid t = 0\}\ , \\ \Sigma_2 = \{(t,x,\tau,\xi) \in \dot{T}^*\tilde{X} \setminus N^*X \mid \tau = \lambda(t,x,\xi)\}. \end{cases}
$$

As a consequence of the results of Chazarain [8, Theorem 2.10] it follows from (i), (ii) and (iii) that there exist p d o's $C_j(t,x,D_t,D_x) \in L^{m-r}(\tilde{X})$ such that we have the decomposition

$$
(3.4)\qquad P \equiv \sum_{j=0}^{r} C_j Q^j \quad , \text{ near } \rho_o\ ,
$$

where $Q = D_t - \lambda(t,x,D_x)$.

As a general remark we observe that without loss of generality we can assume that in the representation (3.1) the operators P_{m-j}, for $j = 0,\dots,k-1$, have full symbols $p_{m-j}(t,x;\tau,\xi)$; this does not change any of the principal symbols at

$t = 0$. Analogously we assume that the operators C_j, $j = 1,\dots,r$ have full symbols $c_j(t,x,\tau,\xi)$ positively homogeneous of degree $m-r$ in (τ,ξ). We now take advantage of the particular structure of the operator (3.1) to refine the decomposition (3.4).

LEMMA 3.1. Case $k \geq r$: there exist pdo's $B_j(t,x,D_t,D_x) \in L^{m-r}(\tilde{X})$, $j = 0,\dots,r$, and $A_{m-r-\ell}(t,x,D_t,D_x) \in L^{m-r-\ell}(\tilde{X})$, $\ell = 1,\dots,k-r$, such that:

$$P \equiv \sum_{j=0}^{r} t^{k-r+j} B_j Q^j + \sum_{\ell=1}^{k-r} t^{k-r-\ell} A_{m-r-\ell}, \quad \text{near } \rho_0. \tag{3.5}$$

Moreover, the principal symbols b_j of B_j and $a_{m-r-\ell}$ of $A_{m-r-\ell}$ satisfy the condition :

$$\begin{cases} b_j(0,x,\lambda(0,x,\xi),\xi) = \dfrac{1}{j!} (\partial_\tau^j p_{m-r+j})(0,x,\lambda(0,x,\xi),\xi), \\ \qquad\qquad j = 0,\dots,r \\ a_{m-r-\ell}(0,x,\tau,\xi) = p_{m-r-\ell}(0,x,\tau,\xi), \quad \ell = 1,\dots,k-r \end{cases} \tag{3.6}$$

in a conic neighborhood of ρ_0.

Case $k < r$; there exist pdo's $B_j(t,x,D_t,D_x) \in L^{m-r}(\tilde{X})$, $j = r-k$, $r-k+1,\dots,r$, and $A_\ell(t,x,D_t,D_x) \in L^{m-r}(\tilde{X})$, $\ell = 0,1,\dots,r-k-1$, such that:

$$P \equiv \sum_{j=r-k}^{r} t^{(k-r)+j} B_j Q^j + \sum_{\ell=0}^{r-k-1} A_\ell Q^\ell, \quad \text{near } \rho_0. \tag{3.7}$$

Moreover, the principal symbol b_j of B_j satisfies the condition :

$$(3.8)\qquad b_j(0,x,\lambda(0,x,\xi),\xi) = \frac{1}{j!}(\partial_\tau^j p_{m-r+j})(0,x,\lambda(0,x,\xi),\xi), \quad j = r-k,\dots,r ,$$

in a conic neighborhood of ρ_o .

Proof. First consider the case $k \geq r$. Comparing the principal symbols in (3.4) and (3.1) we have, near ρ_o, $t^k p_m = t^k e\, q^r = c_r q^r$ (we write $q(t,x,\tau,\xi) = \tau - \lambda(t,x,\xi)$).

Define $B_r \in L^{m-r}(\tilde{X})$ with full symbol $b_r = e$ and write

$$P \equiv \sum_{j=0}^{r} C_j Q^j \equiv t^k B_r Q^r + G_{m-1} , \text{ with } G_{m-1} \equiv \sum_{j=0}^{r-1} C_j Q^j \in L^{m-1}(\tilde{X}) ,$$

since by construction $C_r \equiv t^k B_r$ near ρ_o . Equating the terms of degree $m-1$ in (τ,ξ) we have, in a conic neighborhood of ρ_o :

$$(3.9)\qquad c_{r-1} q^{r-1} + t^k \{b_r \sigma_{r-1}(Q^r) + \frac{1}{i} r q^{r-1} \partial_{t,x} q \cdot \partial_{\tau,\xi} b_r\} = t^{k-1} p_{m-1} ,$$

where $\sigma_{r-1}(Q^r)$ is the term of degree $r-1$ in the symbol of Q^r (incidentally $\sigma_{r-1}(Q^r) = \frac{r(r-1)}{2} q^{r-2} \sigma_1(Q^2)$, for $r \geq 2$).

From (3.9) we obtain $c_{r-1} q^{r-1} = t^{k-1} \tilde{b}_{r-1}$, where

$$(3.10)\qquad \tilde{b}_{r-1} = p_{m-1} - t\{b_r \sigma_{r-1}(Q^r) + \frac{1}{i} r q^{r-1} \partial_{t,x} q \cdot \partial_{\tau,\xi} b_r\} .$$

Note that $\tilde{b}_{r-1}(0,x,\tau,\xi) = p_{m-1}(0,x,\tau,\xi)$.

By Taylor's formula we have

$$c_{r-1}(t,x,\tau,\xi) = \sum_{\ell=0}^{k-2} \frac{1}{\ell!}(\partial_t^\ell c_{r-1})(0,x,\tau,\xi)t^\ell + t^{k-1}\tilde{c}_{r-1}(t,x,\tau,\xi)$$

for some smooth symbol $\tilde{c}_{r-1}$. Since $c_{r-1}\, q^{r-1} = t^{k-1}\tilde{b}_{r-1}$ we obtain $\partial_t^\ell c_{r-1}(0,x,\tau,\xi) = 0$, for $\ell = 0,\dots,k-2$ and $(0,x,\tau,\xi) \notin \Sigma_2$. In conclusion near ρ_o, we have :

$$\tilde{c}_{r-1}\, q^{r-1} = \tilde{b}_{r-1}\ . \tag{3.11}$$

Again, by Taylor's formula :

$$\tilde{b}_{r-1}(t,x,\tau,\xi) = \sum_{\ell=0}^{r-2} \frac{1}{\ell!}(\partial_\tau^\ell \tilde{b}_{r-1})(t,x,\lambda(t,x,\xi),\xi)q^\ell + q^{r-1}\, b_{r-1}(t,x,\tau,\xi),$$

for some symbol b_{r-1}. By comparison with (3.11) we finally obtain

$$\tilde{c}_{r-1} = b_{r-1}\ , \quad \text{near } \rho_o\ . \tag{3.12}$$

Observe that

$$\begin{cases} b_r(0,x,\lambda(0,x,\xi),\xi) = \dfrac{1}{r!}(\partial_\tau^r p_m)(0,x,\lambda(0,x,\xi),\xi)\ , \\ b_{r-1}(0,x,\lambda(0,x,\xi),\xi) = \dfrac{1}{(r-1)!}(\partial_\tau^{r-1} p_{m-1})(0,x,\lambda(0,x,\xi),\xi)\ , \end{cases} \tag{3.13}$$

in a conic neighborhood of ρ_o .

Define $B_{r-1} \in L^{m-r}(\tilde{X})$ with full symbol b_{r-1}. By construction $t^{k-1} B_{r-1} \equiv C_{r-1}$.

Thus we can write $P \equiv \sum_o^r C_j Q^j \equiv t^k B_r Q^r + t^{k-1} B_{r-1} Q^{r-1} + G_{m-2}$, where

$G_{m-2} \equiv \sum_o^{r-2} C_j Q^j \in L^{m-2}(\tilde{X})$.

Equating the terms of degree $m-2$ in (τ,ξ) we have, in a conic neighborhood of ρ_o :

$$c_{r-2}\, q^{r-2} + t^{k-1}\varphi + t^k\psi = t^{k-2} p_{m-2} \quad , \tag{3.14}$$

where φ(resp. ψ) denotes the term of degree $m-2$ in the symbol of $B_{r-1} Q^{r-1}$ (resp. $B_r Q^r$). Thus, near ρ_o we have :

$$c_{r-2}\, q^{r-2} = t^{k-2}\, \tilde{b}_{r-2} \;,\; \tilde{b}_{r-2} = p_{m-2} - t(\varphi + t\psi). \tag{3.15}$$

Note that $\tilde{b}_{r-2}(0,x,\tau,\xi) = p_{m-2}(0,x,\tau,\xi)$.

Proceeding exactly as above we can write :

$$c_{r-2} = t^{k-2}\, b_{r-2} \tag{3.16}$$

in a conic neighborhood of ρ_o , for some smooth symbol b_{r-2} satisfying (3.6) with $j = r-2$. Taking $B_{r-2} \in L^{m-r}(\tilde{X})$ with full symbol b_{r-2} , we have $t^{k-2} B_{r-2} \equiv$ $\equiv C_{r-2}$ near ρ_o . Proceeding in this way we construct pdo's $B_{r-3},\dots,B_1 \in$ $\in L^{m-r}(\tilde{X})$ so that $\sum_{j=1}^{r} t^{k-r+j} B_j Q^j \equiv \sum_1^r C_j Q^j$ and a pdo $B_o \in L^{m-r}(\tilde{X})$

such that $t^{k-r} B_0 = C_0 \bmod L^{m-r-1}(\tilde{X})$; moreover, the principal symbols of B_j are given by (3.6). Equating the terms of degree $m-r-1$ in the above expressions for P yields

$$(3.17) \qquad c_{-1} + \sum_{j=1}^{r} t^{k-r+j} \varphi_j = t^{k-r-1} p_{m-r-1} \text{ , near } \rho_0 \text{ ,}$$

where c_{-1} is the term of degree $m-r-1$ in the symbol of C_0 and φ_j is the term of degree $m-r-1$ in the symbol of $B_j Q^j$. Thus

$$(3.18) \qquad c_{-1} = t^{k-r-1} a_{m-r-1} \text{ , near } \rho_0 \text{ ,}$$

for a smooth symbol a_{m-r-1} satisfying $a_{m-r-1}(0,x,\tau,\xi) = p_{m-r-1}(0,x,\tau,\xi)$, near ρ_0. Defining $A_{m-r-1} \in L^{m-r-1}(\tilde{X})$ with full symbol a_{m-r-1} we obtain that $P \equiv \sum_0^r{}_j \, C_j Q^j \equiv \sum_0^r{}_j \, t^{k-r+j} B_j Q^j + t^{k-r-1} A_{m-r-1} \bmod L^{m-r-2}(\tilde{X})$. Going on we finally obtain the decomposition (3.5), with the principal symbols given by (3.6).

The case $k < r$ is treated similarly (and already follows from the proof above).

q.e.d.

<u>Remark</u>. When $r = 2$ it is easily seen (Cfr. Chazarain [8]) that the Levi condition is satisfied iff :

$$(3.19) \qquad p_{m-1}(t,x,\tau,\xi)\Big|_{\tau=\lambda(t,x,\xi)} = \frac{1}{i}\{t\, e'(t,x,\tau,\xi)(\partial_{t,x} q \cdot \partial_{\tau,\xi} q)\}\Big|_{\tau=\lambda(t,x,\xi)}$$

which amounts to say that the subprincipal symbol

$p_{m-1} - \frac{1}{2i} (\partial^2_{t,\tau} p_m + \sum_{j=1}^{n} \partial^2_{x_j,\xi_j} p_m)$ of the operator P vanishes on Σ_2.

We now briefly recall a well known construction of some Fourier integral operators. Consider the non-linear Cauchy problem:

$$(3.20) \quad \begin{cases} \partial_t \varphi(t,x,\xi) = \lambda(t,x,d_x\varphi(t,x,\xi)), \\ \varphi(0,x,\xi) = \langle x,\xi \rangle, \quad \xi \neq 0. \end{cases}$$

Modify $\lambda(t,x,\cdot)$ out of a neighborhood of $(0,x_0)$ in $\tilde{X}$ to obtain a new smooth $\lambda(t,x,\xi)$ globally defined on $\mathbb{R}^{n+1}_{(t,x)} \times (\mathbb{R}^n_\xi \setminus \{0\})$ which is independent of x for $|x|$ large. For this new λ we know that the Cauchy problem (3.20) has a unique smooth real solution φ defined for $(x,\xi) \in \dot{T}^* \mathbb{R}^n$ and $|t| < T_0 \leq T$, for some $T_0 > 0$. Consider now the F.I.O. defined by the oscillatory integral:

$$(3.21) \quad \begin{aligned} Ef(t,x) = (2\pi)^{-(n+1)} \iiiint e^{i[\varphi(t,x,\xi) - \langle y,\xi\rangle + (t-s)\tau]} \times \\ \times\ e(t,x,\xi) f(s,y)\, ds\, dy\, d\xi\, d\tau, \end{aligned}$$

where $f \in C_0^\infty((-T_0,T_0) \times \mathbb{R}^n_x)$ and the symbol $e(t,x,\xi) \in S^0((-T_0,T_0) \times \mathbb{R}^n \times \mathbb{R}^n)$ is chosen in such a way that:

$$(3.22)\quad \begin{cases} e(0,x,\xi) \sim 1, \\ e^{-i\varphi(t,x,\xi)}\,[(D_t - \lambda(t,x,D_x))\,\{e^{i\varphi(\cdot,\cdot,\xi)}e(\cdot,\xi)\}] \sim 0. \end{cases}$$

Note that $\varphi(t,x,\xi) - \langle y,\xi \rangle + (t-s)\tau$ is a non-degenerate phase function in the open cone of $\dot{T}^*\mathbb{R}^{n+1}$ where $|t| < T_o$ and $\xi \neq 0$.

By the general calculus of WF (Cfr. Hörmander [14]) we obtain :

$$(3.23)\quad \begin{aligned} WF'(E) \subset \Lambda = \{((t,x,\tau + \partial_t\varphi(t,x,\xi), d_x\varphi(t,x,\xi)), \\ (t, d_\xi\varphi(t,x,\xi), \tau, \xi)) \in \dot{T}^*\mathbb{R}^{n+1} \times \dot{T}^*\mathbb{R}^{n+1} \mid |t| < T_o,\ \xi \neq 0\}\ . \end{aligned}$$

Put $\tilde{X}_o = (-T_o, T_o) \times X$; since over $t=0$ the conic lagrangian manifold Λ reduces to the graph of a diffeomorphism, there exist a conic neighborhood $V_{\rho_o} \subset \dot{T}^*\tilde{X}_o \subset \dot{T}^*\mathbb{R}^{n+1}$ of $\rho_o = (0, x_o, \lambda(0, x_o\xi^{(o)}), \xi^{(o)})$, a conic neighborhood $V_{\hat{\rho}_o} \subset \dot{T}^*\tilde{X}_o \subset \dot{T}^*\mathbb{R}^{n+1}$ of $\hat{\rho}_o = (0, x_o, 0, \xi^{(o)})$, and a homogeneous canonical transformation $\chi : V_{\hat{\rho}_o} \longrightarrow V_{\rho_o}$ such that

$$(3.24)\quad \begin{cases} \chi(V_{\hat{\rho}_o} \cap \{t=0\}) = V_{\rho_o} \cap \Sigma_1\ , \\ \chi(V_{\hat{\rho}_o} \cap \{\tau=0\}) = V_{\rho_o} \cap \Sigma_2\ , \\ \Lambda \cap (V_{\rho_o} \times V_{\hat{\rho}_o}) = \{(\chi(\rho),\rho) \mid \rho \in V_{\hat{\rho}_o}\}\ . \end{cases}$$

We can obviously suppose that E is invertible on $V_{\rho_o} \times V_{\hat{\rho}_o}$. Thus, for every $f \in \mathscr{E}'(\tilde{X})$ with $WF(f) \subset V_{\hat{\rho}_o}$ we obtain $Ef \in \mathcal{D}'(\tilde{X})$ with $WF(Ef) = \chi(WF(f))$.

By the very definition we have $Q\,E\,f \equiv E\,D_t f$ and $t\,E\,f \equiv E\,t\,f$ near $(\rho_o, \hat{\rho}_o)$.

Consider now the operators (3.5) (case $k \geq r$) and (3.7) (case $k < r$). We are interested in studying $WF(u)$ near ρ_0 for a distribution $u \in \mathcal{D}'(\tilde{X})$ such that $Pu = f$ and $\rho_0 \notin WF(f)$. Using the intertwining elliptic F.I.O. E constructed above we are thus reduced to study $E^{-1} P E = \hat{P}$ near $\hat{\rho}_0 = (0, x_0, 0, \xi^{(o)})$. From 3.1, it follows that we have to consider two different cases:

1) $k \geq r$:

(3.25)
$$\hat{P} \equiv \sum_{j=0}^{r} t^{k-r+j} \hat{B}_j \partial_t^j + \sum_{\ell=1}^{k-r} t^{k-r-\ell} \hat{A}_{m-r-\ell}, \text{ near } \hat{\rho}_0 ,$$

where $\hat{B}_j(t,x,D_t,D_x) \in L^{m-r}(\mathbb{R}^{n+1})$, $j = 0,\dots,r$ and $\hat{A}_{m-r-\ell} \in L^{m-r-\ell}(\mathbb{R}^{n+1})$, $\ell = 1,\dots,k-r$, with principal symbols $\hat{b}_j$, $\hat{a}_{m-r-\ell}$ satisfying:

(3.26)
$$\begin{cases} \hat{b}_j(0,x_0,0,\xi^{(o)}) = \dfrac{1}{j!} (D_\tau^j p_{m-r+j})(0,x_0,\lambda(0,x_0\xi^{(o)}),\xi^{(o)}), \\ \qquad\qquad j = 0,\dots,r, \\ \hat{a}_{m-r-\ell}(0,x_0,0,\xi^{(o)}) = p_{m-r-\ell}(0,x_0,\lambda(0,x_0,\xi^{(o)}),\xi^{(o)}), \\ \qquad\qquad \ell = 1,\dots,k-r. \end{cases}$$

2) $k < r$:

(3.27)
$$\hat{P} \equiv \sum_{j=r-k}^{r} t^{(k-r)+j} \hat{B}_j \partial_t^j + \sum_{\ell=0}^{r-k-1} \hat{A}_\ell \partial_t^\ell, \text{ near } \hat{\rho}_0 ,$$

where $\hat{B}_j \in L^{m-r}(\mathbb{R}^{n+1})$, $j = r-k,\dots,r$ and $\hat{A}_\ell \in L^{m-r}(\mathbb{R}^{n+1})$, $\ell = 0,\dots,r-k-1$, with the principal symbols $\hat{b}_j$ satisfying:

$$\hat{b}_j(0,x_o,0,\xi^{(o)}) = \frac{1}{j!}(D_\tau^j p_{m-r+j})(0,x_0,\lambda(0,x_o,\xi^{(o)}),\xi^{(o)}),$$

(3.28)

$$j = r-k,\dots,r.$$

From now on we shall consider the operators given by (3.25), (3.27); for convenience we suppress the $\hat{}$.

First we treat the case $k \geq r$ and reduce the equation $Pv = g, P$ given by (3.25), to a Fuchsian system of the type studied in Section 2.

LEMMA 3.2. There exist operators $\tilde{B}_j \in L^{m-r}(\mathbb{R}^{n+1})$, $\tilde{A}_{m-r-\ell} \in L^{m-r-\ell}(\mathbb{R}^{n+1})$, $j = 0,\dots,r; \ell = 1,\dots,k-r$, such that the operator given in (3.25) can be written

(3.29)
$$\begin{cases} P \equiv \sum_{j=0}^{r} \tilde{B}_j\, t^{k-r}\, t^j\, \partial_t^j + \sum_{\ell=1}^{k-r} \tilde{A}_{m-r-\ell}\, t^{k-r-\ell}, & \text{near } \rho_0, \\ \tilde{b}_j\big|_{\tau=t=0} = b_j\big|_{\tau=t=0}, & j=0,\dots,r, \end{cases}$$

where $\tilde{b}_j$ (resp. b_j) is the principal symbol of $\tilde{B}_j$ (resp. B_j).

<u>Proof.</u> First we commute t^j with B_j in (3.25). We claim that for every $B \in L^h(\mathbb{R}^{n+1})$ and for every $j \geq 1$ we can write

$$t^j\, B\, \partial_t^j = B\, t^j\, \partial_t^j + \sum_{k=1}^{j} B^{(j,k)}\, t^{j-k}\, \partial_t^{j-k}, \tag{3.30}$$

for some operators $B^{(j,k)} \in L^h(\mathbb{R}^{n+1})$ with principal symbols $b^{(j,k)}$ for which $b^{(j,k)}\big|_{\tau=0} = 0$. If $j=1$, $t\,B\,\partial_t = B\,t\,\partial_t + [t,B]\,\partial_t$; defining $B^{(1,1)} = [t,B]\,\partial_t$ we are o.k. Supposing that (3.30) holds up to some j, we consider $t^{j+1}\,B\,\partial_t^{j+1} =$

$$B\,t^{j+1}\,\partial_t^{j+1} + [t,B]\,t^j\,\partial_t^{j+1} + \sum_{k=1}^{j} B^{(j,k)}\,t^{j-k+1}\partial_t^{j-k+1} + \sum_{k=1}^{j} [t,B^{(j,k)}]\,t^{j-k}\,\partial_t^{j-k+1}.$$

Since for every $p \geq 1$ we can write $t^p\,\partial_t^{p+1} = \sum_{\ell=1}^{p} \alpha_{p,\ell}\,\partial_t(t^{p-\ell}\,\partial_t^{p-\ell})$ for some constants $\alpha_{p,\ell}$, we conclude that (3.30) holds also for $j+1$. Using (3.30) we can thus rewrite P as:

$$(3.31) \qquad P \equiv \sum_{j=0}^{r} t^{k-r}\,\tilde{\tilde{B}}_j\,t^j\,\partial_t^j + \sum_{\ell=1}^{k-r} t^{k-r-\ell}\,A_{m-r-\ell}\,, \text{ near } \rho_o\,,$$

with some new operators $\tilde{\tilde{B}}_j \in L^{m-r}(\mathbb{R}^{n+1})$ whose principal symbols $\tilde{\tilde{b}}_j$ satisfy $\tilde{\tilde{b}}_j\big|_{\tau=0} = b_j\big|_{\tau=0}$.

It is immediate to verify that $\sum_{\ell=1}^{k-r} t^{k-r-\ell}\,A_{m-r-\ell}$ can be rewritten as $\sum_{\ell=1}^{k-r} \tilde{\tilde{A}}_{m-r-\ell}\,t^{k-r-\ell}$ for some new operators $\tilde{\tilde{A}}_{m-r-\ell} \in L^{m-r-\ell}(\mathbb{R}^{n+1})$. To take care of the other terms in (3.31), we write $t^{k-r}\,\tilde{\tilde{B}}_j = \sum_{\ell=0}^{k-r} \tilde{\tilde{B}}_{j,\ell}\,t^{k-r-\ell}$ for some operators $\tilde{\tilde{B}}_{j,\ell} \in L^{m-r-\ell}(\mathbb{R}^{n+1})$, $\tilde{\tilde{B}}_{j,o} = \tilde{\tilde{B}}_j$. Now for $\ell = 1,\dots,k-r$, $j = 0,\dots,r$, we claim that if B is any operator in $L^{m-r-\ell}(\mathbb{R}^{n+1})$, it is possible to write:

$$(3.32)\qquad B\,t^{k-r-\ell}\,t^j\,\partial_t^j = \sum_{\nu=0}^{j-1} B_{\nu,j,\ell}\,t^{k-r}\,t^{j-1-\nu}\partial_t^{j-1-\nu} + \sum_{\mu=1}^{k-r} C_{\mu,j,\ell}\,t^{k-r-\mu}\,,$$

for some operators $B_{\nu,j,\ell} \in L^{m-r}(\mathbb{R}^{n+1})$, $C_{\mu,j,\ell} \in L^{m-r-\mu}(\mathbb{R}^{n+1})$, and the principal symbol of $B_{\nu,j,\ell}$ vanishes at $\tau = 0$ for every ν.

We prove (3.32) by induction. First we fix $\ell = 1$ and proceed by induction on j. For $j = 0$ there is nothing to prove. Suppose that (3.32) holds for $\ell = 1$ and up to $j-1$. Since $t^j\,\partial_t^j = (t\,\partial_t - (j-1))\;t^{j-1}\,\partial_t^{j-1}$, we have :

$$B\,t^{k-r-1}\,t^j\,\partial_t^j = [B\,t^{k-r}\,\partial_t - (j-1)B\,t^{k-r-1}]\;t^{j-1}\,\partial_t^{j-1}$$

$$= [B\,\partial_t\,t^{k-r} - (j-1+k-r)B\,t^{k-r-1}]\;t^{j-1}\,\partial_t^{j-1}\,.$$

Noting that $B\,\partial_t \in L^{m-r}$, and has a principal symbol vanishing for $\tau = 0$, and applying induction we are done. Thus (3.32) is proved for $\ell = 1$.

Supposing that (3.32) holds up to $\ell - 1$, we again proceed by induction on j. For $j = 0$, $B\,t^{k-r-\ell}$ can be absorbed in the sum $\sum_{\mu=1}^{k-r} C_{\mu,0,\ell}\,t^{k-r-\mu}$. If the formula holds up to $j-1$, we have:

$$B\,t^{k-r-\ell}\,t^j\,\partial_t^j = [B\,t^{k-r-(\ell-1)}\,\partial_t - (j-1)B\,t^{k-r-\ell}]t^{j-1}\,\partial_t^{j-1} =$$

$$= [B\,\partial_t\,t^{k-r-(\ell-1)} - (j+k-r-\ell)B\,t^{k-r-(\ell-1)}]\;t^{j-1}\,\partial_t^{j-1}\,.$$

Since $B\,\partial_t \in L^{m-r-(\ell-1)}$, with principal symbol vanishing for $\tau = 0$, and $B \in L^{m-r-\ell} \subset L^{m-r-(\ell-1)}$, applying the induction we are finished.

q.e.d.

For convenience, we put in evidence the following formula:

$$(3.33)\quad \begin{cases} \text{For } h = 0,1,\dots \;;\; j = 1,2,\dots, \\ t^h\, t^j\, \partial_t^j = (t\,\partial_t - h)(t\,\partial_t - (h+1))\dots(t\,\partial_t - (h+j-1))t^h . \end{cases}$$

Let $\Lambda \in L^1(\mathbb{R}^{n+1})$ be an invertible pdo whose symbol near $\tau = 0$ is given by $(1+|\xi|^2)^{1/2}$.

Thus, using Lemma 3.2 and formula (3.33), we can write:

$$(3.34)\quad \begin{cases} P \equiv \sum_{j=0}^{r} \Lambda^{r-k}\, \tilde{B}_j (t\,\partial_t - (k-r))\dots(t\,\partial_t - (k-r+j-1))(\Lambda t)^{k-r} \\ \qquad + \sum_{\ell=1}^{k-r} \Lambda^{r-k+\ell}\, \tilde{\tilde{A}}_{m-r-\ell} (\Lambda t)^{k-r-\ell} \quad , \text{ near } \rho_0 , \end{cases}$$

where $\tilde{\tilde{A}}_{m-r-\ell}$ are some new operators in $L^{m-r-\ell}(\mathbb{R}^{n+1})$.

For convenience, we suppress in (3.34) the $\sim$, $\approx$ and only keep in mind that the principal symbol b_j of $B_j = \tilde{B}_j$ satisfies

$$(3.35)\quad b_j(\rho_0) = \frac{1}{j!}\,(D_\tau^j\, p_{m-r+j})(0,x_0,\lambda(0,x_0,\xi^{(o)}),\, \xi^{(o)}),\; j=0,\dots,r.$$

Suppose that $Pv = f$, $v \in \mathcal{D}'(\mathbb{R}^{n+1})$, with $WF(v)$ concentrated in a conic neighborhood of ρ_0. Define

$$
(3.36) \quad \begin{cases}
v_1 = v \\
v_2 = \Lambda t v = \Lambda t v_1 \\
\dots\dots\dots\dots\dots\dots \\
v_{k-r+1} = (\Lambda t)^{k-r} v = \Lambda t v_{k-r} \\
v_{k-r+2} = (t \partial_t - (k-r)) v_{k-r+1} \\
v_{k-r+3} = (t \partial_t - (k-r+1)) v_{k-r+2} \\
\dots\dots\dots\dots\dots\dots\dots\dots\dots\dots \\
v_k = (t \partial_t - (k-2)) v_{k-1} .
\end{cases}
$$

Then :

$$
(3.37) \quad \begin{cases}
t \partial_t v_1 = \partial_t (t v_1) - v_1 = - v_1 + \partial_t \Lambda^{-1} v_2 \\
t \partial_t v_2 = \partial_t (t v_2) - v_2 = - v_2 + \partial_t \Lambda^{-1} v_3 \\
\dots\dots\dots\dots\dots\dots\dots\dots\dots\dots \\
t \partial_t v_{k-r} = \partial_t (t v_{k-r}) - v_{k-r} = - v_{k-r} + \partial_t \Lambda^{-1} v_{k-r+1} \\
t \partial_t v_{k-r+1} = (t \partial_t - (k-r)) v_{k-r+1} + (k-r) v_{k-r+1} = (k-r) v_{k-r+} \\
\qquad + v_{k-r+2} \\
t \partial_t v_{k-r+2} = \dots = (k-r+1) v_{k-r+2} + v_{k-r+3} \\
\dots\dots\dots\dots\dots\dots\dots\dots\dots\dots \\
t \partial_t v_{k-1} = \dots = (k-2) v_{k-1} + v_k \\
t \partial_t v_k = (t \partial_t - (k-1)) v_k + (k-1) v_k .
\end{cases}
$$

From (3.30)

$$\Lambda^{r-k} B_r(t\partial_t - (k-r))\dots(t\partial_t - (k-1))(\Lambda t)^{k-r} v =$$

$$= \Lambda^{r-k} B_r(t\partial_t - (k-1))v_k \equiv Pv - \sum_{j=0}^{r-1} \Lambda^{r-k} B_j(t\partial_t - (r-k)) \dots$$

$$\dots (t\partial_t - (r-k+j-1))(\Lambda t)^{k-r} v - \sum_{\ell=1}^{k-r} \Lambda^{r-k+\ell} A_{m-r-\ell}(\Lambda t)^{k-r-\ell} v =$$

$$= f - \sum_{j=0}^{r-1} \Lambda^{r-k} B_j v_{k-r+j+1} - \sum_{\ell=1}^{k-r} \Lambda^{r-k+\ell} A_{m-r-\ell} v_{k-r+\ell+1} .$$

Taking an operator $B_r^{-1} \in L^{-(m-r)}(\mathbb{R}^{n+1})$ such that $B_r^{-1} B_r \equiv I$ near ρ_o, we obtain

$$t\partial_t v_k \equiv (k-1)v_k + B_r^{-1}\Lambda^{k-r} f - \sum_{j=0}^{r-1} B_r^{-1} B_j v_{k-r+j+1}$$

(3.38)

$$- \sum_{\ell=1}^{k-r} B_r^{-1}\Lambda^{\ell} A_{m-r-\ell} v_{k-r+\ell+1} .$$

From (3.37), (3.38), we see that the vector $(v_1,\dots,v_k) = \vec{v}$ satisfies a system:

$$(3.39) \qquad (t\partial_t I_k - \mathcal{B}(t,x,D_t,D_x))\vec{v} = \vec{f} ,$$

where $\vec{f} \in \mathcal{D}'(\mathbb{R}^{n+1})^k$, with $WF(\vec{f}) = WF(f)$ near ρ_o ; and $\mathcal{B} \in L^o(\mathbb{R}^{n+1}; k \times k)$ with principal symbol $\mathcal{B}(t,x,\tau,\xi)$ such that :

(3.40) $B(\rho_o) =$

$$
= \left(\begin{array}{c|cccccc}
-I_{k-r} & & & \square & & & \\
\hline
 & k-r & 1 & & & & \\
 & & k-r+1 & & & & \square \\
\square & & & & & & \\
 & \square & & & & & \\
 & & & & k-2 & & 1 \\
-\frac{a_{m-k}}{b_r}|\xi|^{k-r} \quad \cdots \quad -\frac{a_{m-r-1}}{b_r}|\xi| & \frac{-b_o}{b_r} & \frac{-b_1}{b_r} & \cdots & \frac{-b_{r-2}}{b_r} & & (k-1)-\frac{b_{r-1}}{b_r}
\end{array}\right)_{\rho=\rho_o}
$$

A computation gives:

(3.41) $\det(z\, I_k - B(\rho_o)) = (z+1)^k \det((z-(k-r))I_r - \mathscr{A}_r)$,

where

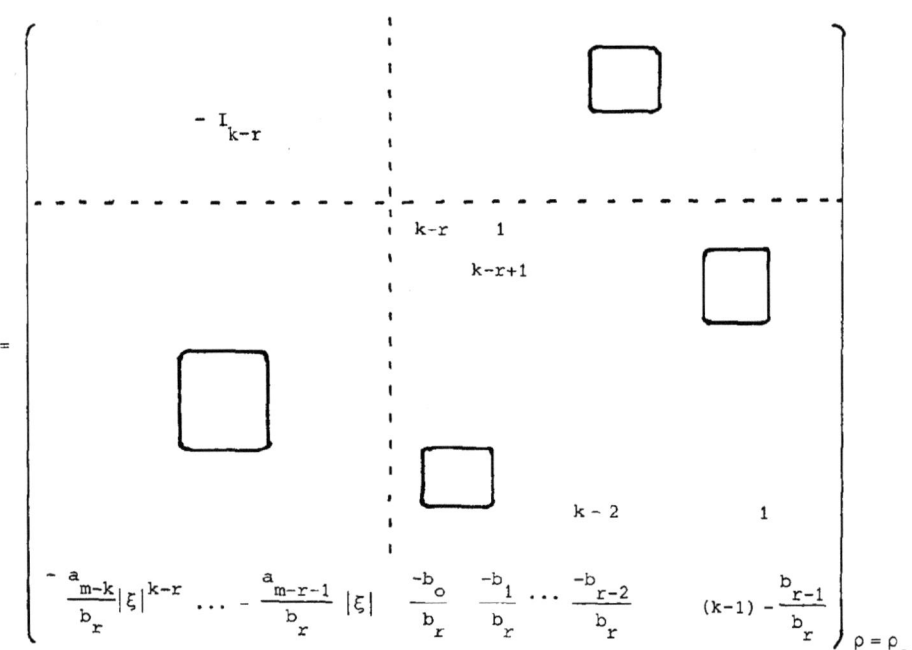

$$
(3.42)\quad \mathscr{A}_r = \left(\begin{array}{cccccc}
0 & 1 & & & & \square \\
 & 1 & 1 & & & \\
 & & 2 & & & \\
 & & & \ddots & & \\
\square & & & & & \\
 & & & & r-2 & 1 \\
-\frac{b_o}{b_r} & -\frac{b_1}{b_r} & \cdots\cdots\cdots & & -\frac{b_{r-2}}{b_r} & (r-1)-\frac{b_{r-1}}{b_r}
\end{array}\right)_{\rho=\rho_o}
$$

and thus:

$$
\det\,(\zeta\, I_r - \mathscr{A}_r) = \sum_{j=0}^{r} \frac{b_j(\rho_o)}{b_r(\rho_o)}\, \zeta(\zeta-1)\ldots(\zeta-(j-1)).
$$

In conclusion:

$$
(3.42)' \quad \begin{aligned}
& b_r(\rho_o)\det(z\, I_k - B(\rho_o)) = \\
& = (z+1)^{k-r} \sum_{j=0}^{r} b_j(\rho_o)(z-(k-r))(z-(k-r+1))\ldots(z-(k-r+j-1))
\end{aligned}
$$

We are now ready to prove the first result of this section. First we give some definitions. Given $\rho_o \in \Sigma_1 \cap \Sigma_2$, denote by $\gamma_1(\rho_o;s)$, $\gamma_2(\rho_o;s)$, $s \in (-\delta,\delta)$ the integral curves in $\dot{T}^*\tilde{X}$ of the Hamiltonian vector fields $H_t = -\frac{\partial}{\partial\tau}$, $H_q = \frac{\partial}{\partial t} - \nabla_\xi\lambda \cdot \frac{\partial}{\partial x} + \frac{\partial\lambda}{\partial t}\frac{\partial}{\partial\tau} + \nabla_x\lambda \cdot \frac{\partial}{\partial\xi}$ passing through ρ_o at $s=0$, and put

$$(3.43)\qquad \begin{cases} \gamma_1^{\pm}(\rho_o) = \{\gamma_1(\rho_o;s) \mid s \in (-\delta,\delta),\ \pm s > 0\} \subset \Sigma_1, \\ \gamma_2^{\pm}(\rho_o) = \{\gamma_2(\rho_o;s) \mid s \in (-\delta,\delta),\ \pm s > 0\} \subset \Sigma_2 \end{cases}$$

(the choice of the orientation has no influence on the subsequent results). If P is given by (3.1) and satisfies hypotheses (i) - (iii) with $k \geqq r$, we define the *indicial polynomial* of P in $\Sigma_o = \Sigma_1 \cap \Sigma_2$ as the map:

$$(3.44)\qquad \begin{cases} I_P : \Sigma_o \times \mathbb{C} \longrightarrow \mathbb{C} \\ I_P(\rho_o;\zeta) = \sum_{j=0}^{r} \frac{1}{j!}(D_\tau^j\, p_{m-r+j})(\hat{\rho}_o)\zeta(\zeta-1)\ldots(\zeta-(j-1)), \\ \rho_o = (0,x_o,\lambda(0,x_o,\xi^{(o)}),\xi^{(o)}),\ \hat{\rho}_o = (0,x_o,\lambda(0,x_o,\frac{\xi^{(o)}}{|\xi^{(o)}|}),\frac{\xi^{(o)}}{|\xi^{(o)}|}),\zeta \in \mathbb{C}. \end{cases}$$

In (3.44) we use the convention that for $j=0$ there is only the term $p_{m-r}(\rho_o)$. Note that the indicial polynomial depends only on $p_m,\ldots,p_{m-r}$.

THEOREM 3.1 Let P be given by (3.1) and suppose that hypotheses (i)—(iii) are satisfied with $k \geqq r$. Let $u \in \mathcal{D}'(\tilde{X})$, $Pu = f$ and let $\rho_o \in \Sigma_o \setminus WF(f)$. Then:

1) If for every $j = 1,2$ and for some choice of the sign +, - we have

$$WF(u) \cap \gamma_j^{\pm}(\rho_o) \cap V = \emptyset \quad ,$$

for some conic neighborhood V of ρ_o, then $\rho_o \notin WF(u)$.

2) Suppose that

$$WF(u) \cap (\gamma_1^{+}(\rho_o) \cup \gamma_1^{-}(\rho_o)) \cap V = \emptyset \ ,$$

for some conic neighborhood V of ρ_o. Then:

i) If $I_P(\rho_o;\zeta) \neq 0$ for $\zeta \in \{-(k-r), -(k-r)+1, (-k-r)+2, \dots\}$, we have $\rho_o \notin WF(u)$.

ii) If $I_P(\rho_o;\zeta) = 0$ for h different values of $\zeta \in \{-(k-r), -(k-r)+1, \dots\}$, there exist distributions $v_1,\dots,v_h \in D'(\tilde{X})$ with

$$WF(v_j) \subset \Sigma_2, \ j = 1,\dots,h \ , \text{ such that } \rho_o \notin WF(u - \sum_{j=1}^{h} v_j).$$

3) Suppose that

$$WF(u) \cap (\gamma_2^{+}(\rho_o) \cup \gamma_2^{-}(\rho_o)) \cap V = \emptyset \ ,$$

for some conic neighborhood V of ρ_o. Then :

i) If $I_P(\rho_o;\zeta) \neq 0$ for $\zeta \in \{-(k-r)-1, -(k-r)-2, \dots\}$, there exists a distribution $u_o \in D'(\tilde{X})$ with $WF(u_o) \subset \Sigma_1$ such that $\rho_o \notin WF(u - u_o)$.

When $k = r$ we can take $u_o = 0$.

ii) If $I_P(\rho_o;\zeta) = 0$ for h different values of $\zeta \in \{-(k-r)-1, -(k-r)-2 \ldots\}$, there exist distributions $w_1,\ldots,w_h \in \mathcal{D}'(\tilde{X})$ with $WF(w_j) \subset \Sigma_1$, $j = 1,\ldots,h$, such that $\rho_o \notin WF(u - u_o - \sum_{j=1}^{h} w_j)$. When $k = r$, we can take $u_o = 0$.

<u>Proof.</u> Using Lemma 3.1 write, microlocally near ρ_o, P as in (3.5). Application of the elliptic F.I.O. E constructed above transforms P into the operator $\hat{P}$ (3.25), noting that the canonical relation associated with P transforms the bicharacteristics of t and τ into the bicharacteristics of t and $q = \tau - \lambda(t,x,\xi)$ respectively. Using Lemma 3.2, reduce $\hat{P}$ to the form (3.29), near ρ_o. Reducing the corresponding equation to the system (3.39) and taking into account that the eigenvalues of the matrix $B(\rho_o)$ are $z = -1$, if $k > r$, and $z = \zeta_j + (k-r)$ if ζ_j denote the roots of the indicial equation $I_P(\rho_o;\zeta) = 0$, $j = 1,\ldots,r$, all the statements in the theorem follow from Theorem 2.2.

q.e.d.

We now turn to consider the case $k < r$ and reduce the equation $Pv = g$, P given by (3.27) to a Fuchsian system of dimension r of the type considered in Section 2. We thus consider:

$$(3.45) \qquad P \equiv \sum_{j=r-k}^{r} t^{k-r+j} B_j \partial_t^j + \sum_{\ell=0}^{r-k-1} A_\ell \partial_t^\ell \quad , \text{ near } \rho_o,$$

where $B_j \in L^{m-r}(\mathbb{R}^{n+1})$, $j=r-k,\dots,r$; $A_\ell \in L^{m-r}(\mathbb{R}^{n+1})$, $\ell = 0,\dots,r-k-1$,

and the principal symbol b_j of B_j satisfies:

(3.46)
$$b_j(0,x_o,0,\xi^{(o)}) = \frac{1}{j!}(D_\tau^j p_{m-r+j})(0,x_o,\lambda(0,x_o,\xi^{(o)}),\xi^{(o)}),$$

$j=r-k,\dots,r.$

We write (3.45) as

$$P \equiv \sum_{j=0}^{k} t^j B_{j+r-k}\,\partial_t^j\,\partial_t^{r-k} + \sum_{\ell=0}^{r-k-1} A_\ell\,\partial_t^\ell \quad ,$$

and using the same arguments as in the proof of Lemma 3.2.:

(3.47)
$$P \equiv \sum_{j=0}^{k} \tilde{B}_{j+r-k}\,t^j\,\partial_t^j\,\partial_t^{r-k} + \sum_{\ell=0}^{r-k-1} A_\ell\,\partial_t^\ell \text{ , near } \rho_o,$$

for some new operators $\tilde{B}_{j+r-k} \in L^{m-r}(\mathbb{R}^{n+1})$ whose principal symbols $\tilde{b}_{j+r-k}$ verify $\tilde{b}_{j+r-k}\big|_{\tau=0} = b_{j+r-k}\big|_{\tau=0}$.

For convenience we suppress the ~, so that :

(3.48)
$$\begin{cases} b_{j+r-k}(0,x_o,0,\xi^{(o)}) = \dfrac{1}{(j+r-k)!}(D_\tau^{j+r-k}\,p_{m+j-k})(0,x_o,\lambda(0,x_o,\xi^{(o)}),\xi^{(o)}) \\ j = 0,\dots,k. \end{cases}$$

Suppose now that $Pv = f$, $v \in \mathcal{D}'(\mathbb{R}^{n+1})$, with $WF(v)$ concentrated in a conic neighborhood of $\rho_o = (0,x_o,0,\xi^{(o)})$. Define :

$$
(3.49)\qquad \begin{cases}
v_1 = v \\
v_2 = \partial_t v_1 = \partial_t v \\
\cdots\cdots\cdots\cdots \\
v_{r-k+1} = \partial_t v_{r-k} = \partial_t^{r-k} v \\
v_{r-k+2} = t\,\partial_t v_{r-k+1} = t\,\partial_t(\partial_t^{r-k} v) \\
v_{r-k+3} = (t\,\partial_t - 1)\, v_{r-k+2} = t^2\,\partial_t^2(\partial_t^{r-k} v) \\
\cdots\cdots\cdots\cdots\cdots\cdots\cdots \\
v_r = (t\,\partial_t - (k-2))v_{r-1} = t^{k-1}\,\partial_t^{k-1}(\partial_t^{r-k} v).
\end{cases}
$$

Then:

$$
(3.50)\qquad \begin{cases}
t\,\partial_t v_1 = t\,v_2 \\
t\,\partial_t v_2 = t\,v_3 \\
\cdots\cdots\cdots\cdots \\
t\,\partial_t v_{r-k} = t\,v_{r-k+1} \\
t\,\partial_t v_{r-k+1} = v_{r-k+2} \\
t\,\partial_t v_{r-k+2} = v_{r-k+2} + v_{r-k+3} \\
t\,\partial_t v_{r-k+3} = 2\,v_{r-k+3} + v_{r-k+4} \\
\cdots\cdots\cdots\cdots\cdots\cdots \\
t\,\partial_t v_{r-1} = (k-2)v_{r-1} + v_r \\
t\,\partial_t v_r = (t\,\partial_t - (k-1))v_r + (k-1)v_r = (k-1)v_r + t^k\,\partial_t^k(\partial_t^{r-k} v).
\end{cases}
$$

Now:

$$t^k\,\partial_t^k(\partial_t^{r-k}\,v) \equiv B_r^{-1}\,P\,v - \sum_{j=0}^{k-1} B_r^{-1}\,B_{j+r-k}\,v_{r-k+j+1} - \sum_{j=0}^{r-k-1} B_r^{-1}\,A_j\,v_{j+1}\,.$$

Thus we obtain a system for the vector $(v_1,\dots,v_r) = \vec{v}$,

$$(t\,\partial_t\,I_r - \mathcal{B}(t,x,D_t,D_x))\vec{v} = \vec{f}\,, \tag{3.51}$$

where $\vec{f} \in \mathcal{D}'(\mathbb{R}^{n+1})^r$, with $WF(\vec{f}) = WF(f)$ near ρ_0 , and $\mathcal{B} \in L^0(\mathbb{R}^{n+1}; r\times r)$ with principal symbol $\mathcal{B}(t,x,\tau,\xi)$ such that :

$$\mathcal{B}(\rho_o) = \left(\begin{array}{c:c} \square_{r-k,r-k} & \square_{r-k,k} \\ \hdashline \square & \begin{matrix} 0 & 1 & & & \\ & 1 & 1 & & \\ & & 2 & 1 & \\ & & & \ddots & \ddots \\ & & & k-2 & 1 \end{matrix} \\ \hdashline -\dfrac{a_o}{b_r}\cdots\cdots -\dfrac{a_{r-k-1}}{b_r} & -\dfrac{b_{r-k}}{b_r}\ \cdots\ -\dfrac{b_{r-2}}{b_r} \quad -\dfrac{b_{r-1}}{b_r}+(k-1) \end{array}\right)_{\rho=\rho_o} \tag{3.52}$$

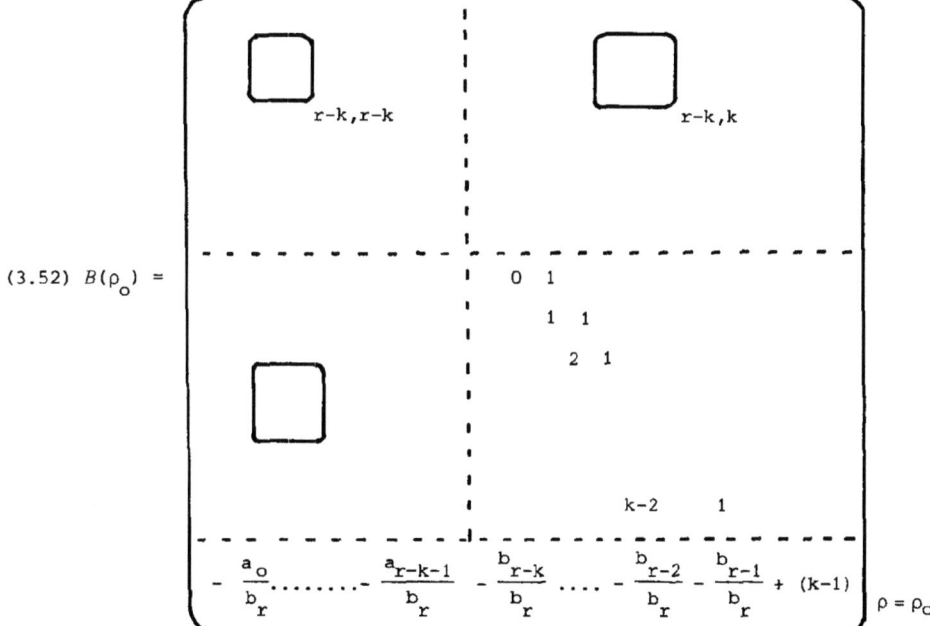

A computation gives:

$$\det(z\, I_r - B(\rho_o)) = z^{r-k} \det(z\, I_k - \mathscr{A}_k), \quad \text{where:}$$

$$(3.53) \quad \mathscr{A}_k = \left(\begin{array}{cccccc} 0 & 1 & & & & \\ & 1 & 1 & & \mathbf{0} & \\ & & 2 & 1 & & \\ & & & \ddots & \ddots & \\ & \mathbf{0} & & & k-2 & 1 \\ -\dfrac{b_{r-k}}{b_r} & \cdots\cdots\cdots & & & -\dfrac{b_{r-2}}{b_r} & -\dfrac{b_{r-1}}{b_r} + (k-1) \end{array} \right)_{\rho = \rho_o}$$

Thus:

$$\det(z\, I_k - \mathscr{A}_k) = \sum_{j=0}^{k} \frac{b_{j+r-k}(\rho_o)}{b_r(\rho_o)} z(z-1)\ldots(z-(j-1)),$$

and therefore :

$$b_r(\rho_o)\det(z\, I_r - B(\rho_o)) =$$

(3.54)

$$= z^{r-k} \sum_{j=0}^{k} b_{j+r-k}(\rho_o) z(z-1)\dots(z-(j-1)),$$

where we use the convention that for $j=0$ there is only the term $b_{r-k}(\rho_o)$. If the operator P is given by (3.1) and satisfies hypotheses (i)——(iii) with $r > k$, we define the *indicial polynomial* of P in $\Sigma_o = \Sigma_1 \cap \Sigma_2$ as the map:

$$(3.55)\quad \begin{cases} I_P : \Sigma_o \times \mathbb{C} \longrightarrow \mathbb{C} \\ I_P(\rho_o;\zeta) = \displaystyle\sum_{j=0}^{k} \frac{1}{(j+r-k)!} (D_\tau^{j+r-k} p_{m+j-k})(\hat{\rho}_o)\zeta(\zeta-1)\dots(\zeta-(j-1)) \\ \rho_o = (0,x_o,\lambda(0,x_o,\xi^{(o)}),\xi^{(o)}),\ \hat{\rho}_o = (0,x_o,\lambda(0,x_o\dfrac{\xi^{(o)}}{|\xi^{(o)}|}),\dfrac{\xi^{(o)}}{|\xi^{(o)}|}),\ \zeta \in \mathbb{C}. \end{cases}$$

We now have the corresponding result:

THEOREM 3.2. Let P be given by (3.1) and suppose that hypotheses (i)——(iii) are satisfied with $r > k$. Let $u \in D'(\tilde{X})$, $Pu = f$ and let $\rho_o \in \Sigma_o \setminus WF(f)$. Then:

1) If for every $j = 1,2$ and for some choice of the sign $+,-$ we have

$$WF(u) \cap (\gamma_j^{\pm}(\rho_o) \cap V = \emptyset \qquad ,$$

for some conic neighborhood V of ρ_o, then $\rho_o \notin WF(u)$.

2) Suppose that

$$WF(u) \cap (\gamma_1^+(\rho_o) \cup \gamma_1^-(\rho_o)) \cap V = \emptyset ,$$

for some conic neighborhood V of ρ_o. Then :

i) If $I_p(\rho_o;\zeta) \neq 0$ for $\zeta \in \{0,1,\dots\}$, there exists a distribution $v_o \in \mathcal{D}'(\tilde{X})$ with $WF(v_o) \subset \Sigma_2$ such that $\rho_o \notin WF(u - v_o)$.

ii) If $I_p(\rho_o;\zeta) = 0$ for h different values of $\zeta \in \{0,1,\dots\}$, there exist distributions $v_1,\dots,v_h \in \mathcal{D}'(\tilde{X})$ with $WF(v_j) \subset \Sigma_2$, $j = 1,\dots,h$, such that $\rho_o \notin WF(u - v_o - \sum_{j=1}^{h} v_j)$.

3) Suppose that

$$WF(u) \cap (\gamma_2^+(\rho_o) \cup \gamma_2^-(\rho_o)) \cap V = \emptyset ,$$

for some conic neighborhood V of ρ_o. Then:

i) If $I_p(\rho_o;\zeta) \neq 0$ for $\zeta \in \{-1,-2,\dots\}$, we have $\rho_o \notin WF(u)$.

ii) If $I_p(\rho_o;\zeta) = 0$ for h different values of $\zeta \in \{-1,-2,\dots\}$, there exist distributions $w_1,\dots,w_h \in \mathcal{D}'(\tilde{X})$ with $WF(w_j) \subset \Sigma_1$, $j = 1,\dots,h$, such that $\rho_o \notin WF(u - \sum_{j=1}^{k} w_j)$.

The proof goes as in Theorem 3.1, taking into account (3.54). q.e.d.

Remarks. A comparison between Theorems 3.1 and 3.2 shows that the case $k = r$ plays a particular role. When $k > r$ (resp. $k < r$) we cannot conclude that

$\rho_o \notin WF(u)$ if we only know that $WF(u) \cap [\gamma_2^+(\rho_o) \cup \gamma_2^-(\rho_o)] = \emptyset$ (resp. that $WF(u) \cap [\gamma_1^+(\rho_o) \cup \gamma_1^-(\rho_o)] = \emptyset$).

Consider, for example, $P = \Delta_{t,x}(t^2\partial_t + t\alpha(t,x))$, where $\Delta_{t,x}$ is the Laplacean in $\mathbb{R}^{n+1}$ and $\alpha \in C^\infty(\mathbb{R}^{n+1})$.

In this case $m = 3$, $k = 2$, $r = 1$ and obviously $P(\delta_t \otimes \varphi(x)) = 0$ for every $\varphi \in \mathcal{D}'(\mathbb{R}^n)$. If $(x_o,\xi^{(o)}) \in WF(\varphi)$, we have $(\gamma_2^+(\rho_o) \cup \gamma_2^-(\rho_o)) \cap WF(u) = \emptyset$, but $\rho_o = (0,x_o,0,\xi^{(o)}) \in WF(u)$. On the other hand consider in $\mathbb{R}^{n+1}$ the operator $P = t\partial_t^2 + \alpha(t)\,\partial_t + \beta(t)$ (so that $m = r = 2$, $k = 1$). If $\alpha(t)$, $\beta(t)$ are analytic in a neighborhood of $t = 0$, the ordinary equation $Pv(t) = 0$ has a non trivial solution $v(t)$ which is analytic in a neighborhood of $t = 0$ provided $-\alpha(0) \notin \{0,1,2,\dots\}$. Thus $u = v(t) \otimes \varphi(x)$ solves $Pu = 0$ for every $\varphi \in \mathcal{D}'(\mathbb{R}^n)$. If $(x_o,\xi^{(o)}) \in WF(\varphi)$, we have $(\gamma_1^+(\rho_o) \cup \gamma_1^-(\rho_o)) \cap WF(u) = \emptyset$, but $\rho_o = (0,x_o,0,\xi^{(o)}) \in WF(u)$.

We now analyse a Fuchsian partial differential operator near the conormal bundle $\dot{N}^*(X)$. For the rest of this section we will consider a differential operator:

$$P = \sum_{j=0}^{k} t^{k-j}\, P_{m-j}(t,x,D_t,D_x), \tag{3.56}$$

where:

$$P_{m-j} = a_{m-j}(t,x)D_t^{m-j} + \sum_{\substack{|\alpha| + \ell \leq m-j \\ \ell < m-j}} a_{j,\alpha\ell}(t,x)D_x^\alpha D_t^\ell\ , \quad j = 0,\dots,k. \tag{3.57}$$

Denoting by p_{m-j} the principal symbol of P_{m-j}, we suppose that :

(3.58) $$p_m \big|_{\dot{N}^*X} \neq 0 \quad (\text{i.e. } a_m(0,x) \neq 0 \,,\quad \forall\, x \in X).$$

The following Lemma will be useful.

LEMMA 3.3. There exist differential operators $Q_{m-j}(t,x,D_t,D_x)$ of order m, $j = 0,1,\dots,k$, for which :

(3.59) $$P = Q_m t^k + Q_{m-1} t^{k-1} + \dots + Q_{m-k}\,.$$

Moreover, for every $\rho = (0,x,\pm 1,0) \in \dot{N}^*X$ and for every $j = 0,\dots,k$:

(3.60) $$\begin{cases} q_{m-j}(0,x,\pm 1,0) = (\pm 1)^{m-j} \displaystyle\sum_{\ell=0}^{j} \frac{(-1)^\ell}{\ell!\, i^\ell}\, c_\ell\, a_{m-j+\ell}(0,x)\,, \quad \text{with} \\ c_\ell = \begin{cases} 1, & \text{if } \ell = 0\,, \\ (m-j+\ell)\dots(m-j+1)(k-j+\ell)\dots(k-j+1), & \text{if } \ell \geq 1\,, \end{cases} \end{cases}$$

where q_{m-j} denotes the principal symbol of Q_{m-j}.

Proof. Formula (3.59) is obvious, since for every j we can write

(3.61) $$t^{k-j} P_{m-j} = P^{(o)}_{m-j} t^{k-j} + P^{(1)}_{m-j} t^{k-j-1} + \dots + P^{(k-j)}_{m-j}$$

for some differential operators $P^{(\ell)}_{m-j}$ of order $m-j-\ell$ $(P^{(o)}_{m-j} = P_{m-j})$. It is then enough to put:

$$(3.62) \quad Q_{m-j} = \begin{cases} P_m = P^{(o)}_m & , \text{ if } j=0 , \\ P^{(o)}_{m-j} + P^{(1)}_{m-(j-1)} + \ldots + P^{(j)}_m & , \text{ if } 1 \leqq j \leqq k . \end{cases}$$

Thus,

$$(3.62)' \quad q_{m-j}(\rho) = \begin{cases} p_m(\rho) & , \text{ if } j=0 , \\ \sum_{\ell=0}^{j} p^{(\ell)}_{m-j+\ell}(\rho) & , \text{ if } 1 \leqq j \leqq k , \end{cases}$$

where $p^{(\ell)}_{m-j+\ell}$ denotes the principal symbol of $P^{(\ell)}_{m-j+\ell}$.

The natural remark to do is that, if we put

$$(3.63) \qquad \hat{P}_{m-j} = a_{m-j}(t,x) D_t^{m-j} \quad , \quad j = 0,1,\ldots,k,$$

and write as above:

$$(3.63)' \qquad t^{k-j} \hat{P}_{m-j} = \hat{P}^{(o)}_{m-j} t^{k-j} + \hat{P}^{(1)}_{m-j} t^{k-j-1} + \ldots + \hat{P}^{(k-j)}_{m-j} ;$$

then the following relations hold :

$$(3.64)\qquad p^{(\ell)}_{m-j+\ell}(0,x,\pm 1,0) = \hat{p}^{(\ell)}_{m-j+\ell}(0,x,\pm 1,0), \quad \left\{ \begin{array}{l} j = 0,\ldots,k, \\ \ell = 0,\ldots,j, \end{array} \right.$$

where $\hat{p}^{(\ell)}_{m-j+\ell}$ is the principal symbol of $\hat{P}^{(\ell)}_{m-j+\ell}$.

To compute $\hat{p}^{(\ell)}_{m-j+\ell}(0,x,\pm 1,0)$ we use the formula :

$$(3.65)\qquad t^p D_t^q = D_t^q t^p + \sum_{j=1}^{p} \frac{(-1)^j}{j!\,i^j} [q(q-1)\ldots(q-j+1)] \cdot p(p-1)\ldots(p-j+1)\ D_t^{q-j}\ t^{p-j},$$

which holds for every p,q with $1 \leq p \leq q$. Formula (3.65) is easily proved by induction.

From (3.65) it follows that for $j = 0,1,\ldots,k$:

$$(3.66)\qquad \hat{P}^{(\ell)}_{m-j+\ell} = \left\{ \begin{array}{ll} a_{m-j}(t,x) & , \text{ if } \ell = 0, \\ \frac{(-1)^\ell}{\ell!\,i^\ell}(m-j+\ell)\ldots(m-j+1)(k-j+\ell)\ldots(k-j+1) & \\ \quad \cdot\ a_{m-j}(t,x)D_t^{m-j} & , \text{ if } \ell \geq 1. \end{array} \right.$$

Then (3.60) follows from (3.62)', (3.64) and (3.66).

q.e.d.

Consider now the operator (3.59) and let $\tilde{Q} \in L^{-m}(\tilde{X})$ be a p d o such that $\tilde{Q}\, Q_m \equiv Q_m\, \tilde{Q} \equiv I$ near $\dot{N}^*X$ (the existence of $\tilde{Q}$ is a consequence of (3.58)). Applying $\tilde{Q}$ on the left in (3.59) and putting $\tilde{Q}\, P = \tilde{P}$, we obtain the operator:

(3.67) $$\tilde{P} \equiv t^k + B_{-1}\, t^{k-1} + \ldots + B_{-k}\ , \text{ near } \dot{N}^*X\ ,$$

where $B_{-j} = \tilde{Q}\, Q_{m-j} \in L^{-j}(\tilde{X})$, $j = 1,\ldots,k$.

The original operator P is microlocally equivalent, near $\dot{N}^*X$, to the operator $\tilde{P}$ in (3.67).

Now let $\Lambda \in L^1(\tilde{X})$ be an invertible operator with full symbol τ in a conic neighborhood of $\dot{N}^*X$.

If we have the equation $\tilde{P}u = t^k u + B_{-1}\, t^{k-1}\, u + \ldots + B_{-k}\, u = f$, $u, f \in D'(\tilde{X})$, define:

(3.68) $$\begin{cases} u_1 = \Lambda^{-(k-1)}\, u \\ u_2 = \Lambda^{-(k-2)}\, t\, u \\ \ldots\ldots\ldots\ldots\ldots \\ u_{k-1} = \Lambda^{-1}\, t^{k-2}\, u \\ u_k = t^{k-1}\, u\, . \end{cases}$$

It is easy to verify that :

$$(3.69)\qquad t\,u_j = [t, \Lambda^{-(k-j)}]\,\Lambda^{k-j}\,u_j + \Lambda^{-1}\,u_{j+1}\,, \qquad j = 1,\dots,k-1.$$

Moreover,

$$(3.70)\qquad \begin{aligned} t\,u_k &= t^k\,u \equiv f - B_{-1}\,t^{k-1}\,u - \dots - B_{-k}\,u = \\ &= f - B_{-k}\,\Lambda^{k-1}\,u_1 - B_{-(k-1)}\,\Lambda^{k-2}\,u_2 - \dots - B_{-1}\,u_k\,. \end{aligned}$$

Thus the operator $\tilde{P}$ is microlocally equivalent near $\dot{N}^*X$ to the system:

$$(3.71)\qquad t\,I_k\,v = B(t,x,D_t,D_x)\,v + g\,,$$

where $v = (v_1,\dots,v_k)$, $g = (0,\dots,0,f)$ and $B \in L^{-1}(\tilde{X};\, k\times k)$. Applying the operator $i\,D_t\,I_K$ on the right in (3.71), system $t\,I_k - B$ is transformed into the system

$$(3.72)\qquad t\,\partial_t\,I_k - A(t,x,D_t,D_x)$$

where $A \in L^{o}(\tilde{X};\, k\times k)$ is given by:

$$
(3.73) \qquad A = \begin{pmatrix}
i[t,D_t^{-(k-1)}]D_t^k & i & & & \Box \\
& i[t,D_t^{-(k-2)}]D_t^{k-1} & i & & \\
& & \ddots & \ddots & \\
\Box & & & i[t,D_t^{-1}]D_t^2 & i \\
-i\,B_{-k}\,D_t^k & -i\,B_{-(k-1)}\,D_t^{k-1} & \cdots & -i\,B_{-2}\,D_t^2 & -i\,B_{-1}\,D_t
\end{pmatrix}
$$

microlocally near $\dot{N}^*X$. The principal symbol $a_o(t,x,\tau,\xi)$ of the operator A satisfies:

$$
(3.74) \quad a_o(0,x,\pm 1,0) = \begin{pmatrix}
k-1 & i & & & \Box \\
& k-2 & i & & \\
& & \ddots & \ddots & \\
\Box & & & 1 & i \\
-i\,b_{-k}(0,x,\pm 1,0)(\pm 1)^k & \cdots & & & -i\,b_{-1}(0,x,\pm 1,0)(\pm 1)
\end{pmatrix}
$$

where b_{-j}, $j = 1,\dots,k$ is the principal symbol of B_{-j}.

A computation shows that the eigenvalues of $a_o(0,x,\pm 1,0)$ are the roots of the polynomial equation:

$$\begin{aligned}(3.75)\qquad & \lambda(\lambda-1)\ \ldots\ (\lambda-k+1)\ + \\ & + i(\lambda-1)\ \ldots\ (\lambda-k+1)\ [b_{-1}(\rho)(\pm 1)]\ + \\ & + i^2(\lambda-2)\ \ldots\ (\lambda-k+1)\ [b_{-2}(\rho)(\pm 1)^2]\ + \ldots + \\ & + i^k[b_{-k}(\rho)(\pm 1)^k] = 0\ ,\quad \rho = (0,x,\pm 1,0).\end{aligned}$$

Taking into account that $b_{-j}(\rho) = q_{m-j}(\rho)/q_m(\rho)$, $j = 1,\ldots,k$, equation (3.75) becomes:

$$\begin{aligned}(3.75)'\qquad & \lambda(\lambda-1)\ \ldots\ (\lambda-k+1)q_m(\rho)\ + \\ & + i(\lambda-1)\ \ldots\ (\lambda-k+1)\ [(\pm 1)q_{m-1}(\rho)]\ + \\ & + \ldots + i^k\ [(\pm 1)^k\ q_{m-k}(\rho)] = 0.\end{aligned}$$

We can rewrite (3.75)' in terms of the $p_{m-j}(\rho) = a_{m-j}(0,x)(\pm 1)^{m-j}$, $j = 0,\ldots,k$, using (3.60). Equation (3.75)' becomes:

$$\begin{aligned}(3.76)\qquad & J_k(\lambda)a_m(0,x) + i\ J_{k-1}(\lambda)a_{m-1}(0,x) + i^2\ J_{k-2}(\lambda)a_{m-2}(0,x) \\ & + \ldots + i^k\ J_o(\lambda)a_{m-k}(0,x) = 0,\end{aligned}$$

where $J_{k-j}(\lambda)$ is a polynomial of degree $k-j$ in λ, given by:

$$J_{k-j}(\lambda) = (\lambda - j) \ldots (\lambda - k + 1) +$$

$$+ \frac{(-1)^1}{1!} (m - j)(k - j)(\lambda - j - 1) \ldots (\lambda - k + 1) +$$

(3.77)
$$+ \frac{(-1)^2}{2!} (m - j)(m - j - 1)(k - j)(k - j - 1)(\lambda - j - 2) \ldots (\lambda - k - 1) +$$

$$+ \ldots + \frac{(-1)^{k-j}}{(k-j)!} (m - j)(m - j - 1) \ldots (m - k + 1)(k - j)! \;, \quad j = 0,\ldots,k \;.$$

A rather messy computation shows that :

(3.78)
$$J_{k-j}(\lambda) = (\lambda - m)(\lambda - 1 - m) \ldots (\lambda - (k - j - 1) - m) \;, \quad j = 0,\ldots,k$$

(convention $J_o(\lambda) = 1$).

In conclusion , defining the *indicial polynomial* of the operator P given in (3.56) (and satisfying (3.58)) as the map:

(3.79)
$$\begin{cases} I_P : X \times \mathbb{C} \longrightarrow \mathbb{C} \\ I_P(x;\lambda) = \lambda(\lambda - 1) \ldots (\lambda - k + 1) a_m(0,x) + \\ \quad + i\lambda(\lambda - 1) \ldots (\lambda - k + 2) a_{m-1}(0,x) + \ldots + i^k a_{m-k}(0,x), \\ x \in X \;,\; \lambda \in \mathbb{C}. \end{cases}$$

we see that the eigenvaluesof the matrix $a_o(0,x,\pm 1,0)$ are exactly the roots of the equation

(3.80) $$I_p(x;\lambda - m) = 0 .$$

It is interesting to note that $\sigma(a_o(0,x,+1,0)) = \sigma(a_o(0,x,-1,0))$ for every $x \in X$. A direct application of Theorem 2.3 yields the following theorem.

THEOREM 3.3. Consider the operator P given by (3.56) which satisfies (3.58). For every $\rho_\pm = (0,x_o, \tau = \pm 1, \xi = 0) \in \dot{N}^*X$ denote by $P_{\rho_\pm} : M_{\rho_\pm}(\tilde{X}) \to M_{\rho_\pm}(\tilde{X})$ the linear map induced by P on the stalk over $\rho_\pm$ of the sheaf $M(\tilde{X})$ of microdistributions on $\tilde{X}$.

Then $\ker P_{\rho_\pm} \simeq D'_{x_o}(X)^k$, Coker $P_{\rho_\pm} = \{0\}$.

Remarks. Using the observations following the proof of Theorem 2.3, we can give a more precise description of $\ker P_{\rho_\pm}$ if we make assumptions on the indicial equation (3.80). Denote by $\lambda_1(x),\dots,\lambda_k(x)$ the roots of the indicial equation (3.80). Suppose that $\lambda_i(x_o) - \lambda_j(x_o) \notin \mathbb{Z}$ for $i \neq j$; moreover, suppose that $\lambda_i(x_o) \notin \mathbb{Z}$, $i = 1, \dots, k$. Then $\ker P_{\rho_\pm}$ is isomorphic to the subspace of microdistributions of the form

(3.81) $$\pi_{\rho_\pm}\left[(t \pm i\, o)^{\lambda_j(x)} \otimes \varphi_j^\pm(x)\right],$$

$j = 1,\dots,k$, for some distributions $\varphi_j^+, \varphi_j^- \in D'_{x_o}(X)$, with $\pi_{\rho_\pm} : M(\tilde{X}) \to M_{\rho_\pm}(\tilde{X})$ being the canonical map.

4. OPERATORS WITH MULTIPLE NON-INVOLUTIVE CHARACTERISTICS

In this Section we apply the results of Sections 2 and 3 to the analysis of the phenomenon of "branching" of the singularities for some operators with multiple characteristics.

Using the notation of Section 1, let $P \in L^m(\tilde{X})$, $m \in \mathbb{R}$, with symbol $p(t,x,\tau,\xi) \sim p_m(t,x,\tau,\xi) + p_{m-1}(t,x,\tau,\xi) + \dots$; we suppose that p_m can be factored, on $\dot{T}^*\tilde{X} \setminus N^*X$ in the following way:

$$(4.1) \qquad p_m(t,x,\tau,\xi) = e(t,x,\tau,\xi)(\tau - \lambda_1(t,x,\xi))^k (\tau - \lambda_2(t,x,\xi))^r ,$$

where $k,r = 1,2,\dots$, $\lambda_j \in C^\infty(\tilde{X} \times (\mathbb{R}^n \setminus 0))$, $j = 1,2$, are real functions positively homogeneous of degree 1 with respect to ξ, and e is a smooth function positively homogeneous of degree $m-(k+r)$ with respect to (τ,ξ). Put $\varphi_j(t,x,\tau,\xi) = \tau - \lambda_j(t,x,\xi)$, $j = 1,2$, and define $\Sigma_j = \varphi_j^{-1}(0) \subset \dot{T}^*\tilde{X} \setminus N^*X$. We suppose that

$$(4.1)' \qquad \begin{cases} \text{i)} & \Sigma_1 \cap \Sigma_2 = \Sigma \neq \emptyset ; \\ \text{ii)} & \{\varphi_1, \varphi_2\} \neq 0 \quad \text{on } \Sigma; \\ \text{iii)} & e\big|_{\Sigma_1 \cup \Sigma_2} \neq 0 . \end{cases}$$

Furthermore, we shall also suppose that the operator P verifies a Levi condition with respect to each multiple factor:

L_1) Let $\rho_1 \in \Sigma_1$, then there exist pdo's $B_j \in L^{m-k}(\tilde{X})$, $j = 0,1,\dots,k$, such that:

$$P(t,x,D_t,D_x) \equiv \sum_{j=0}^{k} B_j(t,x,D_t,D_x)(\varphi_1(t,x,D_t,D_x))^j, \quad \text{near } \rho_1 ;$$

$L_2)$ Let $\rho_2 \in \Sigma_2$, then there exist pdo's $C_j \in L^{m-r}(\tilde{X})$, $j=0,1,\dots,r$, such that

$$P(t,x,D_t,D_x) \equiv \sum_{j=0}^{r} C_j(t,x,D_t,D_x)(\varphi_2(t,x,D_t,D_x))^j, \quad \text{near } \rho_2 .$$

Let now $\rho_o \in \Sigma$; we denote by $\gamma_j(\rho_o;s) : (-\varepsilon,\varepsilon) \subset \mathbb{R} \longrightarrow \Sigma_j$, $j=1,2$, the integral curve of the Hamiltonian vector field H_{φ_j}, satisfying $\gamma_j(\rho_o;0) = \rho_o$ and put $\gamma_j^{\pm}(\rho_o) = \{\gamma_j(\rho_o;s) \mid \pm s \in (0,\varepsilon)\}$, $j=1,2$. From hypothesis ii) it follows that $\gamma_j(\rho_o;s) \in \Sigma_j \setminus \Sigma$ for $s \neq 0$.

Given a distribution $u \in \mathcal{D}'(\tilde{X})$ we are interested in studying $WF(u) \setminus WF(Pu)$ near Σ.

From now on we will denote by ρ_o a fixed point belonging to Σ and we suppose that $\{\varphi_1, \varphi_2\}(\rho_o) < 0$ (the case $\{\varphi_1, \varphi_2\}(\rho_o) > 0$ can be handled quite analogously, interchanging the role of φ_1 and φ_2).

Put $\psi_1 = \{\varphi_2, \varphi_1\}^{-1} \varphi_1$, $\psi_2 = \varphi_2$, so that $\{\psi_1, \psi_2\}(\rho_o) = -1$. Applying the construction of Duistermaat - Sjöstrand [11, Lemma 3.2], one can find two real C^∞-functions α, β defined in a conic neighborhood of ρ_o, which are positively homogeneous of degree 0 and such that:

1) $\alpha - 1$, $\beta - 1$ are flat at Σ;

2) $\{\alpha\psi_1, \beta\psi_2\} = -1$, identically in a conic neighborhood of ρ_o.

Define $q_1 = \alpha\psi_1$, $q_2 = \beta\psi_2$, so that, in a conic neighborhood of ρ_o,

$p_m = \tilde{b}_o \, q_1^k \, q_2^r$, where $\tilde{b}_o = e \, \{\varphi_2, \varphi_1\}^k \alpha^{-k} \beta^{-r}$.

Without loss of generality we may suppose that $\tilde{b}_o$, q_1, q_2 are defined on all of $\dot{T}^*\tilde{X}$ and denote by $Q_j \in L^{j-1}(\tilde{X})$ a p d o with full symbol q_j, $j = 1,2$. In what follows we shall need the following lemma, which states an "invariance property" of the Levi condition.

For convenience we put $y = (t,x), \eta = (\tau,\xi)$.

LEMMA 4.1. Let $P \in L^m(\tilde{X})$, $p \sim p_m + p_{m-1} + \dots$, where

$$p_m(t,x,\tau,\xi) = h(t,x,\tau,\xi)(\ell(t,x,\tau,\xi))^k$$

(here $\ell(t,x,\tau,\xi) = \tau - \lambda(t,x,\xi)$ is a symbol positively homogeneous of degree 1 with respect to (τ,ξ)).

Suppose that P verifies a Levi condition near $\ell^{-1}(0)$, so that, possibly thinking of all symbols as defined on the whole of $\dot{T}^*\tilde{X}$:

$$(4.1) \qquad P \equiv \sum_{j=0}^{k} A_j \, L^j ,$$

where L is a p d o with symbol $\ell(t,x,\tau,\xi)$ near $\ell^{-1}(0)$, $A_j \in L^{m-k}(\tilde{X})$, $j = 0,1,\dots,k$. Let now $B \in L^{-1}(\tilde{X})$ be elliptic in a conic neighborhood of $\ell^{-1}(0)$; denote by $Q \in L^o(\tilde{X})$ the p d o whose symbol is $q = b\,\ell$ (b being the symbol of B).

Thenthere exist p d o's $A_j' \in L^{m-k+j}(\tilde{X})$, $j = 0,\dots,k$, such that

$$(4.2) \qquad P \equiv \sum_{j=0}^{k} A_j' \, Q^j \quad , \text{ near } \ell^{-1}(0) = q^{-1}(0).$$

Proof. Taking the principal symbol in (4.1) we obtain

$$p_m(t,x,\tau,\xi) = a_k(t,x,\tau,\xi)(\ell(t,x,\tau,\xi))^k =$$

$$= (a_k(t,x,\tau,\xi)b^{-k}(t,x,\tau,\xi))(q(t,x,\tau,\xi))^k.$$

We denote by $A'_k \in L^m(\tilde{X})$ a pdo with full symbol $a_k b^{-k} = a'_k$.

Take now the symbol of order $m-1$ of $P - A'_k Q^k$:

$$\sigma_{m-1}(P - A'_k Q^k) = p_{m-1} - \frac{k}{i} q^{k-1} \partial_\eta a'_k \cdot \partial_y q$$

(4.3)

$$- \frac{k(k-1)}{2} a'_k q^{k-2} \sigma_{-1}(Q^2).$$

Now, from (4.1),

$$p_{m-1} = a_{k-1} \ell^{k-1} + \frac{k}{i} \ell^{k-1} \partial_\eta a_k \cdot \partial_y \ell + \frac{k(k-1)}{2} a_k \ell^{k-2} \sigma_1(L^2).$$

Noting that

$$\sigma_{-1}(Q^2) = \frac{1}{i} [\ell^2 \partial_\eta b \cdot \partial_y b + q(\partial_\eta b \cdot \partial_y \ell + \partial_\eta \ell \cdot \partial_y b) + b^2 \partial_\eta \ell \cdot \partial_y \ell]$$

formula (4.3) can be rewritten as

$$\sigma_{m-1}(P - A'_k Q^k) = q^{k-1} \quad [-\frac{k}{i}\partial_\eta a'_k \cdot \partial_y q - \frac{k(k-1)}{2i} a'_k(\partial_\eta b \cdot \partial_y \ell +$$

$$\partial_\eta \ell \cdot \partial_y b) + a_{k-1} b^{-(k-1)} + \frac{k}{i} b^{-(k-1)} \partial_\eta a_k \cdot \partial_y \ell -$$

$$q \frac{k(k-1)}{2i} a'_k b^{-2} \partial_\eta b \cdot \partial_y b] = q^{k-1} \varphi .$$

Hence we may denote by A'_{k-1} the p d o belonging to $L^{m-1}(\tilde{X})$ having symbol φ (homogeneous of degree $m-1$ with respect to η).

This procedure can now be started again, and so on.

q.e.d.

We now use the classical Hamilton-Jacobi theory to find a homogeneous canonical transformation χ from a conic neighborhood $V \subset \dot{T}^*\tilde{X}$ of ρ_o to an open cone $\chi(V) \subset \dot{T}^* \mathbb{R}^{n+1}$ (where the variables are denoted by (s,y,σ,η), $(s,y) \in \mathbb{R}^{n+1}$), so that $\chi(\rho_o) = \hat{\rho}_o = (0,y_o,0,\eta^{(o)})$ and

$$(4.4) \qquad \begin{cases} s \circ \chi = q_1 \\ \sigma \circ \chi = q_2 . \end{cases}$$

We can associate to χ two properly supported Fourier integral operators $F \in I^o(\tilde{X} \times \mathbb{R}^{n+1}; \Gamma')$, $F^{-1} \in I^o(\mathbb{R}^{n+1} \times \tilde{X}; (\Gamma^{-1})')$, where Γ (resp. Γ^{-1}) is a closed conic neighborhood of $(\hat{\rho}_o,\rho_o)$ (resp. $(\rho_o, \hat{\rho}_o)$) contained in the graph of χ (resp. χ^{-1}), such that

$$(4.5)\quad \begin{cases} WF'(F^{-1}F - I_{\tilde{X}}) \not\ni (\rho_o,\rho_o),\ WF'(F\,F^{-1} - I_{\mathbb{R}^{n+1}}) \not\ni (\hat{\rho}_o,\hat{\rho}_o), \\ F\,Q_1\,F^{-1} \in L^o(\mathbb{R}^{n+1}), \quad \text{with principal symbol } s \text{ near } \hat{\rho}_o\,; \\ F\,Q_2\,F^{-1} \in L^1(\mathbb{R}^{n+1}), \quad \text{"} \quad \text{"} \quad \text{"} \quad \sigma \quad \text{"} \quad \hat{\rho}_o\,. \end{cases}$$

Thus $F\,Q_2\,F^{-1} = D_s + A_o$, near $\hat{\rho}_o$, with $A_o \in L^o(\mathbb{R}^{n+1})$.

Using a well known argument (cf. Duistermaat - Hörmander [10]), one can find an elliptic p d o $C \in L^o(\mathbb{R}^{n+1})$ such that $C^{-1}(D_s + A_o)C \equiv D_s$ near $\hat{\rho}_o$. Putting $\tilde{F} = C^{-1}F$, $\tilde{F}^{-1} = F^{-1}C$, we have

$$(4.6)\quad \begin{cases} \tilde{F}\,Q_2\,\tilde{F}^{-1} \equiv D_s\,, \quad \text{near } \hat{\rho}_o\,, \\ \tilde{F}\,Q_1\,\tilde{F}^{-1} \equiv s + A_{-1}\,, \quad \text{near } \hat{\rho}_o, \quad \text{for some } A_{-1} \in L^{-1}(\mathbb{R}^{n+1}). \end{cases}$$

Moreover the F.I.O.'s $\tilde{F}$, $\tilde{F}^{-1}$ still have the property $(\rho_o,\rho_o) \notin WF'(\tilde{F}^{-1}\tilde{F} - I_{\tilde{X}})$, $(\hat{\rho}_o,\hat{\rho}_o) \notin WF'(\tilde{F}\,\tilde{F}^{-1} - I_{\mathbb{R}^{n+1}})$.

Now, keeping in mind the definition of Q_1,Q_2, by Lemma 4.1, we get that the operator P satisfies two (partial) Levi conditions with respect to each factor Q_1,Q_2; using the F.I.O., $\tilde{F}$ and $\tilde{F}^{-1}$ as intertwining operators (and denoting again the new variables by (t,x,τ,ξ)) it is easy to see that hypotheses $L_1)$ and $L_2)$ are transformed into the following :

$L_1)$ Let $\tilde{\Sigma}_1 = \{(0,x,\tau,\xi) \mid (\tau,\xi) \neq (0,0)\}$ and $\hat{\rho}_1 = (0,x_o,\tau^o,\xi^o) \in \tilde{\Sigma}_1$. Then there are p d o's $\tilde{B}_j \in L^{m-(k-j)}(\mathbb{R}^{n+1})$ $j = 0,1,\dots,k$, such that

$$(4.7)\qquad \tilde{P} \equiv \sum_{j=0}^{k} \tilde{B}_j\, t^j\,, \quad \text{near } \hat{\rho}_1\,.$$

L_2) Let $\tilde{\Sigma}_2 = \{(t,x,0,\xi) \mid \xi \neq 0\}$ and $\hat{\rho}_2 = (t_o,x_o,0,\xi^o) \in \tilde{\Sigma}_2$. Then there are p d o's $\tilde{C}_j \in L^{m-r}(\mathbb{R}^{n+1})$, $j = 0,1,\dots,r$, such that

(4.8) $$\tilde{P} \equiv \sum_{j=0}^{r} \tilde{C}_j D_t^j \ , \quad \text{near} \quad \hat{\rho}_2 \ .$$

We now observe that commuting $\tilde{B}_j$ with t^j in (4.7) yields the representation:

(4.9) $$\tilde{P} \equiv \sum_{j=0}^{k} t^{k-j} \tilde{P}_{m-j} \ , \quad \text{near} \quad \hat{\rho}_1 \ ,$$

for some new p d o's $\tilde{P}_{m-j} \in L^{m-j}(\mathbb{R}^{n+1})$, $j = 0,\dots,k$.

Now we are exactly in the same situation as in Section 3, taking into account that $\tilde{P}$ satisfies a Levi condition with respect to the factor τ^r .

As a trivial consequence of Theorems 3.1 and 3.2 we have the following (partial) result.

THEOREM 4.1. Let $P \in L^m(\tilde{X})$ be a p d o satisfying (4.1), (4.1)' and the Levi conditions L_1), L_2). Let $\rho_o \in \Sigma$ and let $u \in \mathcal{D}'(\tilde{X})$ be a distribution for which $\rho_o \notin WF(Pu)$.

If for every $j = 1,2$ we have $\gamma_j^{\pm}(\rho_o) \cap WF(u) = \emptyset$ for some choice of the sign + or − , then $\rho_o \notin WF(u)$.

To obtain results on $WF(u)$ knowing only that $(\gamma_j^+(\rho_o) \cup \gamma_j^-(\rho_o)) \cap WF(u) = \emptyset$ for a $j \in \{1,2\}$ requires a knowledge of the "indicial polynomial" of P at ρ_o. To express the indicial polynomial in terms of the full symbol of the operator P does not seem to be an easy task. More precisely, it is not clear at all how to relate the indicial polynomial of $\tilde{P}$ at $\hat{\rho}_o$ (which can be constructed as in Section 3) with the original operator P !

To give an example we shall explicitly write down the "indicial polynomial" of P in the cases k, $r \leq 2$.

For convenience we put $y = (t,x)$, $\eta = (\tau,\xi)$. The following preparation result for the operator P holds.

LEMMA 4.2.

1. Case $k = r = 1$.

There exist $B_o, B_1 \in L^{m-1}(\tilde{X})$ such that :

$$P \equiv B_o Q_1 Q_2 + B_1 \,, \quad \text{near } \rho_o . \tag{4.10}$$

Moreover, in a conic neighborhood of ρ_o, we have :

$$(4.10)' \quad \begin{cases} b_o|_\Sigma = (e\,\{\varphi_2,\varphi_1\})|_\Sigma \\ b_1|_\Sigma = s_P|_\Sigma + \frac{1}{2i}\,(e\,\{\varphi_2,\varphi_1\})|_\Sigma \end{cases}$$

where b_ℓ is the principal symbol of the operator B_ℓ, $\ell = 0,1$, and s_P denotes the subprincipal symbol of P.

2. Case $k = 1$, $r = 2$.

There exist $B_o, B_1, B_2 \in L^{m-2}(\tilde{X})$ such that

$$P \equiv B_o Q_1 Q_2^2 + B_1 Q_2 + B_2 \,, \quad \text{near } \rho_o . \tag{4.11}$$

Moreover, in a conic neighborhood of ρ_o, we have :

$$
(4.11)' \quad \begin{cases} b_o\big|_\Sigma = (e\,\{\varphi_2,\varphi_1\})\big|_\Sigma \,, \\[2ex] b_1\big|_\Sigma = (\partial_\tau\, p_{m-1})\big|_\Sigma - \frac{1}{i}\, e\big|_\Sigma \,[\partial_\eta\, \varphi_2 \cdot \partial_y\, \varphi_2 + 2\, \partial_\eta\, \varphi_2 \cdot \partial_y\, \varphi_1]_\Sigma \end{cases}
$$

where b_ℓ is the principal symbol of the operator $B_\ell, \ell = 0,1$.

3. Case $k = 2,\ r = 1$.

There exist $B_o, B_1 \in L^{m-1}(\tilde{X})$, $B_2 \in L^{m-2}(\tilde{X})$, such that

$$
(4.12) \qquad P \equiv B_o\, Q_1^2\, Q_2 + B_1\, Q_1 + B_2 \,, \quad \text{near } \rho_o \,.
$$

Moreover, in a conic neighborhood of ρ_o, we have :

$$
(4.12)' \quad \begin{cases} b_o\big|_\Sigma = (e\,\{\varphi_2,\varphi_1\}^2)\big|_\Sigma \,, \\[2ex] b_1\big|_\Sigma = \{\varphi_2,\varphi_1\}\big|_\Sigma \,[\partial_\tau\, p_{m-1} - \frac{1}{i}\, e(\partial_\eta\, \varphi_1 \cdot \partial_y\, \varphi_1 + 2\, \partial_\eta \varphi_1 \cdot \partial_y \varphi_2)]_\Sigma , \end{cases}
$$

where b_ℓ is the principal symbol of the operator B_ℓ, $\ell = 0,1$.

4. Case $k = r = 2$.

There exist $B_o, B_1, B_2 \in L^{m-2}(\tilde{X})$, such that

$$
(4.13) \qquad P \equiv B_o\, Q_1^2\, Q_2^2 + B_1\, Q_1\, Q_2 + B_2 \,, \quad \text{near } \rho_o \,.
$$

Moreover, in a conic neighborhood of ρ_o , we have :

$$(4.13)'\begin{cases} b_o\big|_\Sigma = (e\,\{\varphi_2,\varphi_1\}^2)\big|_\Sigma\ , \\ b_1\big|_\Sigma = \{\varphi_2,\varphi_1\}\big|_\Sigma\ [\frac{1}{2}\,\partial^2_{\tau\tau}\, p_{m-1} - \frac{1}{i}\, e(\partial_\eta\,\varphi_1 \cdot \partial_y\,\varphi_1 + \\ \qquad + \partial_\eta\,\varphi_2 \cdot \partial_y\,\varphi_2 + 4\,\partial_\eta\,\varphi_1 \cdot \partial_y\,\varphi_2]_\Sigma\ , \\ b_2\big|_\Sigma = p_{m-2}\big|_\Sigma - \frac{1}{2i}\,\partial^2_{\tau\tau}\, p_{m-1}\big|_\Sigma\ (\partial_\eta\,\varphi_1 \cdot \partial_y\,\varphi_2)\big|_\Sigma \\ - e\big|_\Sigma\ [(\partial_\eta\,\varphi_1 \cdot \partial_y\,\varphi_2)\{\partial_\eta\,\varphi_1 \cdot \partial_y\,\varphi_1 + 2\,\partial_\eta\varphi_1 \cdot \partial_y\,\varphi_2 + \partial_\eta\,\varphi_2 \cdot \partial_y\,\varphi_2\} \\ \qquad - (\partial_\eta\,\varphi_1 \cdot \partial_y\,\varphi_1)(\partial_\eta\,\varphi_2 \cdot \partial_y\,\varphi_2)]_\Sigma \end{cases}$$

where b_ℓ is the principal symbol of the operator B_ℓ, $\ell = 0,1,2$.

__Proof.__ Define $B_o \in L^{m-r}(\tilde{X})$ with symbol $b_o = e\ \{\varphi_2,\varphi_1\}^k/(\alpha^k\,\beta^r)$ in a conic neighborhood of ρ_o . It follows that the principal symbol of $P - B_o\,Q_1^k\,Q_2^r$ vanishes near ρ_o . The term of order $m-1$ in the symbol of $P - B_o\,Q_1^k\,Q_2^r$ is given by:

$$(4.14)\begin{cases} \sigma_{m-1}(P - B_o\,Q_1^k\,Q_2^r) = \\ = p_{m-1} - [\sigma_{m-r-1}(B_o\,Q_1^k)q_2^r + b_o\,q_1^k\,\sigma_{r-1}(Q_2^r) + \\ \qquad + \frac{1}{i}\,\partial_\eta(b_o\,q_1^k)\cdot\partial_y(q_2^r)]\ , \end{cases}$$

where $\sigma_\ell(\cdot)$ denotes the term of degree ℓ in the symbol of the operator considered. Now:

$$\sigma_{m-r-1}(B_o\, Q_1^k) = \begin{cases} \frac{1}{i}\, \partial_\eta\, b_o \cdot \partial_y\, q_1 \;, & \text{if} \quad k=1, \\ \frac{1}{i}\, b_o\, \partial_\eta\, q_1 \cdot \partial_y\, q_1 + \frac{2}{i}\, q_1\, \partial_\eta\, b_o \cdot \partial_y\, q_1 \;, & \text{if} \quad k=2\;, \end{cases}$$

(4.15)

$$\sigma_{r-1}(Q_2^r) = \begin{cases} 0 \;, & \text{if} \quad r=1, \\ \frac{1}{i}\, \partial_\eta\, q_2 \cdot \partial_y\, q_2 \;, & \text{if} \quad r=2. \end{cases}$$

Using (4.14), (4.15) and the hypotheses on P we now proceed to prove the Lemma.

1. Case $k=r=1$. From (4.14) and (4.15):

$$\sigma_{m-1}(P - B_o\, Q_1\, Q_2) = p_{m-1} - \frac{1}{i}\,[q_2\, \partial_\eta\, b_o \cdot \partial_y\, q_1 + q_1\, \partial_\eta\, b_o \cdot \partial_y\, q_2 + b_o\, \partial_\eta\, q_1 \cdot \partial_y\, q_2]\;. \tag{4.16}$$

Take $B_1 \in L^{m-1}(\tilde{X})$ with symbol $\sigma_{m-1}(P - B_o\, Q_1\, Q_2)$ near ρ_o.

Thus (4.10) is proved. Moreover, using (4.16) we have near ρ_o :

$b_o|_\Sigma = (e\,\{\varphi_2,\varphi_1\})|_\Sigma$, $b_1|_\Sigma = p_{m-1}|_\Sigma - \frac{1}{i}\, b_o|_\Sigma\, (\partial_\eta\, q_1 \cdot \partial_y\, q_2)|_\Sigma =$ $= p_{m-1}|_\Sigma - \frac{1}{i}\, e|_\Sigma\, (\partial_\eta\, \varphi_1 \cdot \partial_y\, \varphi_2)|_\Sigma$. Since $s_P|_\Sigma = p_{m-1}|_\Sigma$ $- \frac{1}{2i}\, b_o|_\Sigma\, [\partial_\eta\, q_1 \cdot \partial_y\, q_2 + \partial_y\, q_1 \cdot \partial_\eta\, q_2]_\Sigma = p_{m-1}|_\Sigma - \frac{1}{2i}\, e|_\Sigma\, \{\varphi_2,\varphi_1\}|_\Sigma$ $\times\, [\partial_\eta\, \varphi_1 \cdot \partial_y\, \varphi_2 + \partial_y\, \varphi_1 \cdot \partial_\eta\, \varphi_2]_\Sigma = b_1|_\Sigma - \frac{1}{2i}\, e|_\Sigma\, \{\varphi_2,\varphi_1\}|_\Sigma$, formula (4.10)' is proved.

2. Case $k = 1$, $r = 2$. From (4.14) and (4.15):

$$\sigma_{m-1}(P - B_o Q_1 Q_2^2) = p_{m-1} - \frac{1}{i}[(\partial_\eta b_o \cdot \partial_y q_1)q_2^2 + b_o q_1(\partial_\eta q_2 \cdot \partial_y q_2)$$

(4.17)

$$+ 2 q_1 q_2(\partial_\eta b_o \cdot \partial_y q_2) + 2 b_o q_2(\partial_\eta b_o \cdot \partial_y q_2)], \quad \text{near } \rho_o .$$

Since s_P vanishes on Σ_2, we obtain near ρ_o :

$$s_P\big|_{\Sigma_2} = p_{m-1}\big|_{\Sigma_2} - \frac{1}{i} b_o q_1(\partial_\eta q_2 \cdot \partial_y q_2)\big|_{\Sigma_2} = 0 .$$

Thus there exists a smooth function $\tilde{c}$, positively homogeneous of degree $m-2$, such that in a conic neighborhood of ρ_o we have :

$$p_{m-1} - \frac{1}{i} b_o q_1(\partial_\eta q_2 \cdot \partial_y q_2) = q_2 \tilde{c} .$$

Using Taylor's formula :

$$\tilde{c}_{|\Sigma} = \partial_\tau p_{m-1|\Sigma} - \frac{1}{i}(b_o(\partial_\eta q_2 \cdot \partial_y q_2))_{|\Sigma}(\partial_\tau q_{1|\Sigma}) \quad (\text{near } \rho_o).$$

Since $\partial_\tau q_{1|\Sigma} = \frac{1}{\{\varphi_2,\varphi_1\}_{|\Sigma}}$ (recall that $\partial_\tau \varphi_1 = 1$), we obtain :

(4.18) $$\tilde{c}_{|\Sigma} = \partial_\tau p_{m-1|\Sigma} - \frac{1}{i}[e(\partial_\eta \varphi_2 \cdot \partial_y \varphi_2)]_\Sigma , \quad \text{near } \rho_o .$$

In conclusion, from (4.17) we get :

$$\sigma_{m-1}(P - B_o\, Q_1\, Q_2^2) =$$

(4.19) $$q_2\,[\,\tilde{c} - \frac{1}{i}\,\{(\partial_\eta\, b_o \cdot \partial_y\, q_1)q_2 + 2(\partial_\eta\, b_o \cdot \partial_y\, q_2)q_1 + 2(\partial_\eta\, q_2 \cdot \partial_y\, q_1)b_o\}]$$

$$= q_2\, b_1$$

Take $B_1 \in L^{m-2}(\tilde{X})$ with symbol b_1 near ρ_o ; formula (4.11) is proved. From (4.18) and (4.19) it follows that :

$$b_1|_\Sigma = \tilde{c}|_\Sigma - \frac{2}{i}\,(e\,\{\varphi_2,\varphi_1\})|_\Sigma\,(\partial_\eta\, q_2 \cdot \partial_y\, q_1)|_\Sigma =$$

$$= \partial_\tau\, p_{m-1}|_\Sigma - \frac{1}{i}\, e|_\Sigma [\partial_\eta\, \varphi_2 \cdot \partial_y\, \varphi_2 + 2\,\partial_\eta\, \varphi_2 \cdot \partial_y\, \varphi_1]_\Sigma \;,$$

so that (4.11)' holds.

3. Case $k = 2$, $r = 1$. From (4.14) and (4.15):

(4.20)
$$\sigma_{m-1}(P - B_o\, Q_1^2\, Q_2) = p_{m-1} - \frac{1}{i}\,[b_o\, q_2(\partial_\eta\, q_1 \cdot \partial_y\, q_1)$$
$$+ 2\, q_1\, q_2(\partial_\eta\, b_o \cdot \partial_y\, q_1) + 2\, b_o(\partial_\eta\, q_1 \cdot \partial_y\, q_2) + q_1(\partial_\eta b_o \cdot \partial_y\, q_2)],\ \text{near } \rho_o.$$

Since S_P vanishes on Σ_1 , we obtain near ρ_o :

$$S_P|_{\Sigma_1} = p_{m-1}|_{\Sigma_1} - \frac{1}{i}\, b_o\, q_2(\partial_\eta\, q_1 \cdot \partial_y\, q_1)|_{\Sigma_1} = 0\ .$$

Thus there exists a smooth function $\tilde{c}$, positively homogeneous of degree $m - 1$, such that in a conic neighborhood of ρ_o we have $\theta = p_{m-1} - \frac{1}{i}\, b_o\, q_2(\partial_\eta\, q_1 \cdot \partial_y\, q_1) = \tilde{c}\, q_1$. Using Taylor's formula :

$$\tilde{c}|_\Sigma = \{\varphi_2,\varphi_1\}|_\Sigma \ \frac{\partial\theta}{\partial\tau}\Big|_{\tau=\lambda_1=\lambda_2} .$$

Therefore :

$$\tilde{c}|_\Sigma = (\partial_\tau p_{m-1}|_\Sigma - \frac{1}{i} e|_\Sigma \ \{\varphi_2,\varphi_1\}^2_\Sigma \left(\frac{\partial_\eta\varphi_1 \cdot \partial_y \varphi_1}{\{\varphi_2 \ \varphi_1\}^2} \right)\Big|_\Sigma \ \{\varphi_2,\varphi_1\}|_\Sigma$$

i.e.,

$$\tilde{c}|_\Sigma = \{\varphi_2,\varphi_1\}|_\Sigma \ [\partial_\tau p_{m-1} - \frac{1}{i} e(\partial_\eta \varphi_1 \cdot \partial_y \varphi_1)]_\Sigma \ , \text{ near } \rho_o \tag{4.21}$$

In conclusion, from (4.20) we get :

$$\begin{aligned} \sigma_{m-1}(P - B_o Q_1^2 Q_2) &= q_1 \ [\tilde{c} - \frac{1}{i} \{2 q_2(\partial_\eta b_o \cdot \partial_y q_1) \\ &+ 2 b_o(\partial_\eta q_1 \cdot \partial_y q_2) + q_1(\partial_\eta b_o \cdot \partial_y q_2)] = q_1 b_1 . \end{aligned} \tag{4.22}$$

Take $B_1 \in L^{m-1}(\tilde{X})$ with symbol b_1 near ρ_o; formula (4.12) holds.

From (4.22) and (4.21) it follows that :

$$\begin{aligned} b_1|_\Sigma &= \tilde{c}|_\Sigma - \frac{2}{i} (e \ \{\varphi_2,\varphi_1\}^2)|_\Sigma \ \left(\frac{\partial_\eta \varphi_1 \cdot \partial_y \varphi_2}{\{\varphi_1,\varphi_2\}} \right)\Big|_\Sigma \\ &= \{\varphi_2,\varphi_1\}|_\Sigma \ [\partial_\tau p_{m-1} - \frac{1}{i} e(\partial_\eta \varphi_1 \cdot \partial_y\varphi_1 + 2 \ \partial_\eta\varphi_1 \cdot \partial_y \varphi_2)]_\Sigma \ , \end{aligned}$$

so that (4.12)' is proved.

4. Case $k = r = 2$. From (4.14) and (4.15):

$$\sigma_{m-1}(P - B_o Q_1^2 Q_2^2) = p_{m-1} - \frac{1}{i} [b_o q_2^2 (\partial_\eta q_1 \cdot \partial_y q_1) +$$

$$(4.23) \qquad + 2 q_1 q_2^2 (\partial_\eta b_o \cdot \partial_y q_1) + b_o q_1^2 (\partial_\eta q_2 \cdot \partial_y q_2) +$$

$$+ 2 q_1^2 q_2 (\partial_\eta b_o \cdot \partial_y q_2) + 4 b_o q_1 q_2 (\partial_\eta q_1 \cdot \partial_y q_2)] \text{, near } \rho_o.$$

Since $S_P|_{\Sigma_1} = 0$, there exists a smooth function $\tilde{c}$, positively homogeneous of degree $m-1$, such that in a conic neighborhood of ρ_o we have :

$$(4.24) \qquad p_{m-1} - \frac{1}{i} b_o q_2^2 (\partial_\eta q_1 \cdot \partial_y q_1) = \tilde{c} q_1 .$$

In the same way, since $S_P|_{\Sigma_2} = 0$, there exists a smooth function $\tilde{d}$, positively homogeneous of degree $m-2$, such that in a conic neighborhood of ρ_o we have :

$$(4.25) \qquad p_{m-1} - \frac{1}{i} b_o q_1^2 (\partial_\eta q_2 \cdot \partial_y q_2) = \tilde{d} q_2 .$$

Thus, near ρ_o :

$$p_{m-1} - \frac{1}{i} [b_o q_1^2 (\partial_\eta q_2 \cdot \partial_y q_2) + b_o q_2^2 (\partial_\eta q_1 \cdot \partial_y q_1)] =$$

$$(4.26) \qquad = q_1 [\tilde{c} - \frac{1}{i} b_o q_1 (\partial_\eta q_2 \cdot \partial_y q_2)] =$$

$$= q_2 [\tilde{d} - \frac{1}{i} b_o q_2 (\partial_\eta q_1 \cdot \partial_y q_1)] .$$

Since q_1 and q_2 are indipendent near ρ_o, there exist smooth functions $\hat{c},\hat{d}$, positively homogeneous of degree $m-2$, such that in a conic neighborhood of ρ_o we have :

$$\begin{aligned} &\tilde{c} - \frac{1}{i} b_o q_1(\partial_\eta q_2 \cdot \partial_y q_2) = q_2 \hat{c}, \\ (4.27) \quad & \\ &\tilde{d} - \frac{1}{i} b_o q_2(\partial_\eta q_1 \cdot \partial_y q_1) = q_1 \hat{d}. \end{aligned}$$

Since $q_1 q_2 \hat{c} = q_1 q_2 \hat{d}$; we obtain that $\hat{c} = \hat{d}$ in a conic neighborhood of ρ_o. Thus (4.23) can be rewritten as :

$$\begin{aligned} &\sigma_{m-1}(P - B_o Q_1^2 Q_2^2) = q_1 q_2 [\hat{c} - \frac{1}{i} \{2 q_2(\partial_\eta b_o \cdot \partial_y q_1) + \\ (4.28) \quad & \\ &+ 2 q_1(\partial_\eta b_o \cdot \partial_y q_2) + 4 b_o(\partial_\eta q_1 \cdot \partial_y q_2)\}] = q_1 q_2 b_1. \end{aligned}$$

Take $B_1 \in L^{m-2}(\tilde{X})$ with symbol b_1 near ρ_o. It follows that the terms of degree m and $m-1$ in the symbol of $P-(B_o Q_1^2 Q_2^2 + B_1 Q_1 Q_2)$ vanish in a conic neighborhood of ρ_o. Thus, there exists $B_2 \in L^{m-2}(X)$ such that

$P \equiv B_o Q_1^2 Q_2^2 + B_1 Q_1 Q_2 + B_2$, near ρ_o.

We now compute the restriction to Σ(near ρ_o) of the principal symbols of the operator B_1 and B_2.

Observe that $b_o|_\Sigma = e|_\Sigma (\{\varphi_2,\varphi_1\}^2)|_\Sigma$, near ρ_o.

To compute $\hat{c}|_\Sigma$ near ρ_o put $\theta = p_{m-1} - \frac{1}{i} b_o q_2^2(\partial_\eta q_1 \cdot \partial_y q_1)$.

We use Taylor's formula:

(4.29) $$\theta = \left(\int_0^1 \left(\frac{\partial\theta}{\partial\tau} \right)(t,x,\lambda_1 + s(\tau - \lambda_1),\xi)\, ds \right) \frac{\{\varphi_2,\varphi_1\}}{\alpha} q_1 ,$$

so that, near ρ_o, we have :

(4.30) $$\tilde{c} = \left[\int_0^1 \left(\frac{\partial\theta}{\partial\tau} \right)(t,x,\lambda_1(t,x,\xi) + s(\tau - \lambda_1(t,x,\xi)),\xi)\, ds \right] \frac{\{\varphi_2,\varphi_1\}}{\alpha}$$

Define :

(4.31) $$\tilde{c} - \frac{1}{i} b_o(\partial_\eta q_2 \cdot \partial_y q_2) q_1 = \omega = \hat{c}\, q_2$$

Using Taylor's formula once more :

(4.32) $$\hat{c} = \left[\int_0^1 \left(\frac{\partial\omega}{\partial\tau} \right)(t,x,\lambda_2(t,x,\xi) + \sigma(\tau - \lambda_2(t,x,\xi)),\xi)\, d\sigma \right] \frac{1}{\beta} .$$

Thus :

(4.33) $$\hat{c}_{|\Sigma} = \left(\frac{\partial\omega}{\partial\tau} \right)(t,x,\lambda_2(t,x,\xi) = \lambda_1(t,x,\xi),\xi)\ \frac{1}{\beta_{|\Sigma}} .$$

Since

$$\frac{\partial\omega}{\partial\tau} = \left(\int_0^1 s\left(\frac{\partial^2\theta}{\partial\tau^2}\right)(t,x,\lambda_1 + s(\tau-\lambda_1),\xi)\,ds \right) \frac{\{\varphi_2,\varphi_1\}}{\alpha}$$

$$+ \left(\int_0^1 \left(\frac{\partial\theta}{\partial\tau}\right)(t,x,\lambda_1 + s(\tau-\lambda_1),\xi)\,ds \right) \partial_\tau \left(\frac{\{\varphi_2,\varphi_1\}}{\alpha} \right)$$

(4.34)

$$- \frac{1}{i}\,\partial_\tau b_0\, q_1 (\partial_\eta q_2 \cdot \partial_y q_2) - \frac{1}{i}\, b_0\, \partial_\tau q_1 (\partial_\eta q_2 \cdot \partial_y q_2)$$

$$- \frac{1}{i}\, b_0\, q_1 \frac{\partial}{\partial\tau} (\partial_\eta q_2 \cdot \partial_y q_2) \;,$$

it follows, from (4.34) and (4.33):

$$\hat{c}\big|_\Sigma = \frac{1}{2}\left(\frac{\partial^2\theta}{\partial\tau^2}\right)(t,x,\lambda_1 = \lambda_2,\xi)\ \{\varphi_2,\varphi_1\}\big|_\Sigma$$

(4.35)
$$+ \left(\frac{\partial\theta}{\partial\tau}\right)(t,x,\lambda_1 = \lambda_2,\xi)\ \left(\frac{\partial}{\partial\tau}\ \{\varphi_2,\varphi_1\}\right)\big|_\Sigma$$

$$- \frac{1}{i}\, b_0\big|_\Sigma\ \frac{\partial q_1}{\partial\tau}\Big|_\Sigma\ (\partial_\eta q_2 \cdot \partial_y q_2)\big|_\Sigma \;.$$

Noting that $\partial_\tau \{\varphi_2,\varphi_1\} = 0$, we conclude that :

$$\hat{c}\big|_\Sigma = \frac{1}{2}\,\partial^2_{\tau\tau}\ p_{m-1}\big|_\Sigma - \frac{1}{i}\, b_0\big|_\Sigma \left(\frac{\partial q_2}{\partial\tau}\right)^2\Big|_\Sigma$$

(4.36)
$$\times(\partial_\eta q_1 \cdot \partial_y q_1)\big|_\Sigma\ \{\varphi_2,\varphi_1\}\big|_\Sigma - \frac{1}{i}\, b_0\big|_\Sigma\ \frac{\partial q_1}{\partial\tau}\Big|_\Sigma (\partial_\eta q_2 \cdot \partial_y q_2)\big|_\Sigma$$

$$= \left[\frac{1}{2}\,\partial^2_{\tau\tau}\ p_{m-1} - \frac{1}{i}\, e(\partial_\eta \varphi_1 \cdot \partial_y \varphi_1 + \partial_\eta \varphi_2 \cdot \partial_y \varphi_2)\right]_\Sigma\ \{\varphi_2,\varphi_1\}\big|_\Sigma \;.$$

From (4.28) we finally obtain:

$$b_1|_\Sigma = \hat{c}|_\Sigma - \frac{1}{i}\, 4\, b_0|_\Sigma\, (\partial_\eta\, q_1 \cdot \partial_y\, q_2)|_\Sigma$$

$$= \hat{c}|_\Sigma - \frac{4}{i}\, e|_\Sigma\, (\partial_\eta\, \varphi_1 \cdot \partial_y\, \varphi_2)|_\Sigma \quad \{\varphi_2,\varphi_1\}|_\Sigma \quad .$$

Thus, using (4.36), the first two formulas in (4.13)' are established. To compute $b_2|_\Sigma$, we observe that we must have

(4.37) $$\sigma_{m-2}(P - B_0\, Q_1^2\, Q_2^2)|_\Sigma - \sigma_{m-2}(B_1\, Q_1\, Q_2)|_\Sigma = b_2|_\Sigma \text{ , near } \rho_0.$$

Now:

$$\sigma_{m-2}(B_1\, Q_1\, Q_2) = \sigma_{m-3}(B_1\, Q_1)q_2 + \frac{1}{i}\, \partial_\eta(b_1\, q_1) \cdot \partial_y\, q_2$$

$$= \frac{1}{i}\, (\partial_\eta\, b_1 \cdot \partial_y\, q_1)q_2 + \frac{1}{i}(\partial_\eta\, q_1 \cdot \partial_y\, q_2)b_1 + \frac{1}{i}\, (\partial_\eta\, b_1 \cdot \partial_y\, q_2)q_1 .$$

Thus, near ρ_o :

$$\sigma_{m-2}(B_1\, Q_1\, Q_2)|_\Sigma = \frac{1}{i}\, b_1|_\Sigma\, (\partial_\eta\, q_1 \cdot \partial_y\, q_2)|_\Sigma$$

(4.38) $$= \frac{1}{i}\, b_1|_\Sigma \left(\frac{\partial_\eta\, \varphi_1 \cdot \partial_y\, \varphi_2}{\{\varphi_2,\varphi_1\}} \right)\Big|_\Sigma \quad .$$

Furthermore :

$$\sigma_{m-2}(B_o\, Q_1^2\, Q_2^2) = \sigma_{m-4}(B_o\, Q_1^2) q_2^2 + b_o\, q_1^2\, \sigma_o(Q_2^2)$$

$$+ \sigma_{m-3}(B_o\, Q_1^2)\sigma_1(Q_2^2) - \frac{1}{2} \sum_{|\alpha|=2} \partial_\eta^\alpha(b_o\, q_1^2)\partial_y^\alpha(q_2^2)$$

$$+ \frac{1}{i}\,\partial_\eta(\sigma_{m-3}(B_o\, Q_1^2)) \cdot \partial_y(q_2^2) + \frac{1}{i}\,\partial_\eta(b_o\, q_1^2) \cdot \partial_y(\sigma_1(Q_2^2)).$$

Taking the restriction to Σ near ρ_o :

$$\sigma_{m-2}(B_o\, Q_1^2\, Q_2^2)|_\Sigma = \sigma_{m-3}(B_o\, Q_1^2)\sigma_1(Q_2^2)|_\Sigma - \frac{1}{2} \sum_{|\alpha|=2} \partial_\eta^\alpha(b_o\, q_1^2)\partial_y^\alpha(q_2^2)|_\Sigma =$$

$$(b_o\, \sigma_{-1}(Q_1^2)\sigma_1(Q_2^2))|_\Sigma - \frac{1}{2} \sum_{|\alpha|=2} \partial_\eta^\alpha(b_o\, q_1^2)\partial_y^\alpha(q_2^2)|_\Sigma =$$

$$- e|_\Sigma\, (\partial_\eta\varphi_1 \cdot \partial_y\, \varphi_1)|_\Sigma\, (\partial_\eta\varphi_2 \cdot \partial_y\, \varphi_2)|_\Sigma$$

$$- 2e|_\Sigma \sum_1^{n+1}{}_{i,j} (\partial_{\eta_i}\, \varphi_1\, \partial_{\eta_j}\, \varphi_1\, \partial_{y_i}\, \varphi_2\, \partial_{y_j}\, \varphi_2)|_\Sigma .$$

In conclusion :

$$\sigma_{m-2}(P - B_o\, Q_1^2\, Q_2^2 - B_1\, Q_1\, Q_2)|_\Sigma = b_2|_\Sigma =$$

(4.39)
$$p_{m-2}|_\Sigma - \Big\{ - e|_\Sigma\, (\partial_\eta\, \varphi_1 \cdot \partial_y\, \varphi_1)(\partial_\eta\, \varphi_2 \cdot \partial_y\, \varphi_2)|_\Sigma$$

$$- 2e|_\Sigma \sum_1^{n+1}{}_{i,j} (\partial_{\eta_i}\, \varphi_1\, \partial_{\eta_j}\, \varphi_1\, \partial_{y_i}\, \varphi_2\, \partial_{y_j}\, \varphi_2)|_\Sigma + \frac{1}{i}\, b_1|_\Sigma \left(\frac{\partial_\eta\, \varphi_1 \cdot \partial_y\, \varphi_2}{\{\varphi_2, \varphi_1\}} \right)\Big|_\Sigma \Big\}$$

Inserting $b_1|_\Sigma$ as given by (4.13)' we easily obtain the last formula in (4.13)!

q.e.d.

Using the Fourier operators $\tilde{F}$ and $\tilde{F}^{-1}$ as intertwining operators in (4.10)-(4.13), it is easily verified that we are reduced to consider near $\hat{\rho}_o$ the following operators

(4.40) <u>Case $k=r=1$:</u>

$$\tilde{P}=\tilde{B}_o s D_s + \tilde{B}_1 \quad , \quad \tilde{B}_o, \tilde{B}_1 \in L^{m-1}(\mathbb{R}^{n+1}),$$

$$\begin{cases} \tilde{b}_o(\hat{\rho}_o) = e(\rho_o)\ \{\varphi_2,\varphi_1\}\ (\rho_o), \\ \tilde{b}_1(\hat{\rho}_o) = s_p(\rho_o) + \frac{1}{2i} e(\rho_o)\ \{\varphi_2,\varphi_1\}\ (\rho_o). \end{cases}$$

(4.41) <u>Case $k=1,\ r=2$:</u>

$$\tilde{P}=\tilde{B}_o s D_s^2 + \tilde{B}_1 D_s + \tilde{B}_2 \ , \quad \tilde{B}_o, \tilde{B}_1, \tilde{B}_2 \in L^{m-2}(\mathbb{R}^{n+1}),$$

$$\begin{cases} \tilde{b}_o(\hat{\rho}_o) = e(\rho_o)\ \{\varphi_2,\varphi_1\}\ (\rho_o), \\ \tilde{b}_1(\hat{\rho}_o) = (\partial_\tau\, p_{m-1})\ (\rho_o) - \frac{1}{i} e(\rho_o)[\partial_\eta\varphi_2 \cdot \partial_y\, \varphi_2 + 2\ \partial_\eta\, \varphi_2 \cdot \partial_y\, \varphi_1]\ (\rho_o). \end{cases}$$

(4.42) <u>Case $k=2,\ r=1$:</u>

$$\tilde{P}=\tilde{B}_o\, s^2\, D_s + \tilde{B}_1\, s + \tilde{B}_2 \ , \quad \tilde{B}_o, \tilde{B}_1 \in L^{m-1}(\mathbb{R}^{n+1}),\ \tilde{B}_2 \in L^{m-2}(\mathbb{R}^{n+1}),$$

$$\begin{cases} \tilde{b}_o(\hat{\rho}_o) = e(\rho_o)\quad \{\varphi_2,\varphi_1\}^2(\rho_o), \\ \tilde{b}_1(\hat{\rho}_o) = \{\varphi_2,\varphi_1\}(\rho_o)[(\partial_\tau\, p_{m-1})(\rho_o) - \frac{1}{i}\, e(\rho_o)\{\partial_\eta\, \varphi_1 \cdot \partial_y\, \varphi_1 + \\ \qquad\qquad + 2\,\partial_\eta\varphi_1 \cdot \partial_y\, \varphi_2]\ (\rho_o). \end{cases}$$

(4.43) Case $k = r = 2$:

$$\tilde{P} = \tilde{B}_o\, s^2 D_s^2 + \tilde{B}_1\, s D_s + \tilde{B}_2\ ,\quad \tilde{B}_o, \tilde{B}_1, \tilde{B}_2 \in L^{m-2}(\mathbb{R}^{n+1}),$$

$$\begin{cases} \tilde{b}_o(\hat{\rho}_o) = e(\rho_o)\ \{\varphi_2,\varphi_1\}^2\ (\rho_o), \\ b_1(\hat{\rho}_o) = \{\varphi_2,\varphi_1\}(\rho_o)\ [(\partial^2_{\tau\tau}\ p_{m-1})(\rho_o) - \frac{1}{i}\, e(\rho_o)\{\partial_\eta\, \varphi_1 \cdot \partial_y\, \varphi_1 + \\ \qquad \partial_\eta\, \varphi_2 \cdot \partial_y\, \varphi_2 + 4\ \partial_\eta\, \varphi_1 \cdot \partial_y\, \varphi_2\}(\rho_o)], \\ \tilde{b}_2(\hat{\rho}_o) = p_{m-2}(\rho_o) - \frac{1}{2i}\, (\partial^2_{\tau\tau}\ p_{m-1})(\rho_o)\ (\partial_\eta\, \varphi_1 \cdot \partial_y\, \varphi_2)(\rho_o) \\ \qquad - e(\rho_o)\ [(\partial_\eta\, \varphi_1 \cdot \partial_y\, \varphi_2)(\rho_o)\ \{\partial_\eta\, \varphi_1 \cdot \partial_y\, \varphi_1 + 2\ \partial_\eta\, \varphi_1 \cdot \partial_y\, \varphi_2 + \\ \qquad \partial_\eta\, \varphi_2 \cdot \partial_y\, \varphi_2\}\ (\rho_o) - (\partial_\eta\, \varphi_1 \cdot \partial_y\, \varphi_1)(\rho_o)(\partial_\eta\, \varphi_2 \cdot \partial_y\, \varphi_2)(\rho_o)]. \end{cases}$$

We will now reduce near $\hat{\rho}_o$ the four operators above to the corresponding Fuchsian systems of the type studied in Section 2 and leave to the reader to state the related propagation results.

1. Case $k = r = 1$.

Denoting by $\tilde{B}_o^{-1} \in L^{-(m-1)}(\mathbb{R}^{n+1})$ a parametrix of $\tilde{B}_o$ near $\hat{\rho}_o$, we are reduced to consider the operator

$$\text{(4.44)} \qquad \hat{P} = s\,\frac{\partial}{\partial s} - \hat{B}_o(s,y,D_s,D_y),$$

where $\hat{B}_o \in L^o(\mathbb{R}^{n+1})$ and its principal symbol at $\hat{\rho}_o$ is given by

$$\text{(4.45)} \qquad \hat{b}_o(\hat{\rho}_o) = \frac{1}{i}\ \frac{s_P(\rho_o)}{e(\rho_o)\{\varphi_2,\varphi_1\}(\rho_o)} - \frac{1}{2}\ ,$$

which is exactly the invariant considered by N. Hanges in [13].

2. Case $k=1$, $r=2$.

Using again $\tilde{B}_o^{-1}$, we are reduced to study the operator

$$(4.46)\qquad \hat{P} = s\frac{\partial^2}{\partial s^2} + \hat{B}_o\frac{\partial}{\partial s} + \hat{B}_1 ,$$

where $\hat{B}_o, \hat{B}_1 \in L^o(\mathbb{R}^{n+1})$ and the principal symbol of $\hat{B}_o$ at $\hat{\rho}_o$ is given by:

$$(4.47)\qquad \hat{b}_o(\hat{\rho}_o) = \frac{1}{i}\ \frac{\partial_\tau\, p_{m-1}(\rho_o)}{e(\rho_o)\ \{\varphi_2,\varphi_1\}(\rho_o)} - \left(\frac{\partial_\eta\varphi_2\cdot\partial_y\varphi_2 + 2\,\partial_\eta\varphi_2\cdot\partial_y\varphi_1}{\{\varphi_2,\varphi_1\}}\right)(\rho_o)$$

If we have the equation $\hat{P}v = f$, define $v_1 = v$, $v_2 = \frac{\partial}{\partial s}v_1$, so that

$$s\frac{\partial}{\partial s}\begin{pmatrix} v_1 \\ v_2 \end{pmatrix} = \begin{pmatrix} 0 & s \\ -\hat{B}_1 & -\hat{B}_o \end{pmatrix}\begin{pmatrix} v_1 \\ v_2 \end{pmatrix} + \begin{pmatrix} 0 \\ f \end{pmatrix}.$$

The eigenvalues of the principal symbol of the matrix $\begin{pmatrix} 0 & s \\ -\hat{B}_1 & -\hat{B}_o \end{pmatrix}$

at $\hat{\rho}_o$ are $\lambda = 0$, $\lambda = -\hat{b}_o(\hat{\rho}_o)$.

3. Case $k=2$, $r=1$.

Using again $\tilde{B}_o^{-1}$ we are reduced to consider the operator

$$(4.48)\qquad \hat{P} = s^2\frac{\partial}{\partial s} + \hat{B}_o s + \hat{B}_1 ,$$

where $\hat{B}_o \in L^o(\mathbb{R}^{n+1})$, $\hat{B}_1 \in L^{-1}(\mathbb{R}^{n+1})$, and the principal symbol of $\hat{B}_o$ at $\hat{\rho}_o$ is given by :

$$(4.49) \qquad \hat{b}_o(\hat{\rho}_o) = i\,\frac{(\partial_\tau p_{m-1})(\rho_o)}{e(\rho_o)\{\varphi_2,\varphi_1\}(\rho_o)} - \left(\frac{\partial_\eta \varphi_1 \cdot \partial_y \varphi_1 + 2\,\partial_\eta \varphi_1 \cdot \partial_y \varphi_2}{\{\varphi_2,\varphi_1\}}\right)(\rho_o).$$

Denote by $\Lambda \in L^1(\mathbb{R}^{n+1})$ an invertible pdo with symbol $(1+|\eta|^2)^{1/2}$ near $\sigma = 0$. If we have the equation $\hat{P}v = f$, define $v_1 = v$, $v_2 = \Lambda s v$, so that

$$s\frac{\partial}{\partial s}\begin{pmatrix} v_1 \\ v_2 \end{pmatrix} = \begin{bmatrix} -1 & \frac{\partial}{\partial s}\Lambda^{-1} \\ -\Lambda\hat{B}_1 & -\Lambda(\hat{B}_o - I)\Lambda^{-1} + [s\frac{\partial}{\partial s}, \Lambda]\,\Lambda^{-1} \end{bmatrix}\begin{pmatrix} v_1 \\ v_2 \end{pmatrix} + \begin{pmatrix} 0 \\ \Lambda f \end{pmatrix}$$

The eigenvalues of the principal symbol of the matrix operator at $\hat{\rho}_o$ are $\lambda = -1$, $\lambda = -\hat{b}_o(\hat{\rho}_o) + 1$.

4. <u>Case k = r = 2.</u>

Using $\tilde{B}_o^{-1}$ again we are reduced to consider the operator

$$(4.50) \qquad \hat{P} = s^2\frac{\partial^2}{\partial s^2} + \hat{B}_o\, s\frac{\partial}{\partial s} + \hat{B}_1\,,$$

where $\hat{B}_o$, $\hat{B}_1 \in L^o(\mathbb{R}^{n+1})$ and their principal symbols at $\hat{\rho}_o$ are given by:

$$\hat{b}_0(\hat{\rho}_0) = i\,\frac{(\partial^2_{\tau\tau}\,p_{m-1}(\rho_0)}{e(\rho_0)\{\varphi_2,\varphi_1\}(\rho_o)} - \left\{\frac{\partial_\eta\varphi_1\cdot\partial_y\varphi_1 + \partial_\eta\varphi_2\cdot\partial_y\varphi_2 + 4\,\partial_\eta\varphi_1\cdot\partial_y\varphi_2}{\{\varphi_2,\varphi_1\}}\right\}(\rho_0)$$

$$\hat{b}_1(\hat{\rho}_0) = -\,\frac{p_{m-2}(\rho_0)}{e(\rho_0)\{\varphi_2,\varphi_1\}^2(\rho_0)} + \frac{1}{2i}\,\frac{(\partial^2_{\tau\tau}\,p_{m-1})(\rho_0)}{e(\rho_0)\{\varphi_2,\varphi_1\}(\rho_0)}$$

(4.51)

$$\times\quad \frac{\partial_\eta\varphi_1\cdot\partial_y\varphi_2}{\{\varphi_2,\varphi_1\}}\,(\rho_o)\; +$$

$$+\left[\frac{(\partial_\eta\varphi_1\cdot\partial_y\varphi_2)\,[\partial_\eta\varphi_1\cdot\partial_y\varphi_1 + 2\partial_\eta\varphi_1\cdot\partial_y\varphi_2 + \partial_\eta\varphi_2\cdot\partial_y\varphi_2] - (\partial_\eta\varphi_1\cdot\partial_y\varphi_1)(\partial_\eta\varphi_2\cdot\partial_y\varphi_2)}{\{\varphi_2,\varphi_1\}^2}\right]$$

If we have the equation $\hat{P}v = f$, define $v_1 = v$, $v_2 = s\frac{\partial}{\partial s}v$ so that:

$$s\frac{\partial}{\partial s}\begin{pmatrix} v_1 \\ v_2 \end{pmatrix} = \begin{pmatrix} 0 & I \\ -\hat{B}_1 & -(\hat{B}_0 - I) \end{pmatrix}\begin{pmatrix} v_1 \\ v_2 \end{pmatrix} + \begin{pmatrix} 0 \\ f \end{pmatrix}.$$

The eigenvalues of the principal symbol of the matrix operator are the roots of the polynomial :

(4.52) $$\lambda^2 + \lambda(\hat{b}_0(\hat{\rho}_0) - 1) + \hat{b}_1(\rho_0).$$

R E F E R E N C E S

[1] ALINHAC S. *Parametrix pour un système hyperbolique à multiplicité variable;* Comm. P.D.E. 2 (3) (1977), 251-296.

[2] ALINHAC S. *Systèmes hyperboliques singuliers;* Astérisque 19 (1974), 3-24

[3] ALINHAC S. *Parametrix et propagation des singularités pour un problème de Cauchy à multiplicité variable;* Astérisque 34-37 (1976), 3-26.

[4] AMANO K. *Branching of singularities for degenerate hyperbolic operators and Stokes phenomena;* Proc. Japan Acad. 56 (1980), 206-209.

[5] BAIOCCHI C., BAOUENDI M.S. *Singular evolution equations;* J. of Funct. Anal. 25 (1977), 103-120.

[6] BAOUENDI M.S., GOULAOUIC C. *Cauchy problem with characteristic initial hypersurface;* Comm. Pure Appl. Math. 26 (1973), 455-475.

[7] BOLLEY P., CAMUS J. *Sur une classe d'opérateurs elliptiques et dégénérés à plusieurs variables;* Bull. Soc. Math. de France, Mémoir 34 (1973), 55-150.

[8] CHAZARAIN J. *Opérateurs hyperboliques à caractéristiques de multiplicité constante;* Ann. Inst. Fourier 24 (1974), 173-202.

[9] CHAZARAIN J. *Propagation des singularités pour une classe d'opérateurs à caractéristiques multiples et résolubilité locale;* Ann. Inst. Fourier 24 (1974), 203-223.

[10] DUISTERMAAT J.J., HÖRMANDER L. *Fourier integral operators II;* Acta Math. 128 (1972), 183-269.

[11] DUISTERMAAT J.J., SJÖSTRAND J. *A global construction for pseudo-differential operators with non-involutive characteristics;* Invent. Math. 20 (1973), 209-225.

[12] HANGES N. *Parametrices and local solvability for a class of singular hyperbolic operators;* Comm. P.D.E. 3 (2) (1978), 105-152.

[13] HANGES N. *Parametrices and propagation of singularities for operators with non-involutive characteristics;* Indiana Univ. Math. J. 28 (1979), 86-97.

[14] HÖRMANDER L. *Fourier integral operators I;* Acta Math. 127 (1971), 79-183.

[15] IVRII V.YA. *Wave fronts of solutions of certain pseudo-differential equations;* Functional Anal. Appl. 10 (1976), 141-142.

[16] IVRII V.YA. *Wave fronts of solutions of certain pseudodifferential equations;* Trans. Moscow Math. Soc. 1 (1981), 49-86.

[17] IVRII V.YA. *Wave fronts of solutions of certain hperbolic pseudodifferential equations;* Trans. Moscow Math. Soc. 1 (1981), 87-119.

[18] KASHIWARA M., KAWAI T., OSHIMA T. *Structure of cohomology groups whose coefficients are microfunctions solutions sheaves of systems of pseudo-differential equations with multiple characteristics,I;* Proc. Japan Acad. 50 (1974), 420-425.

[19] KASHIWARA M., OSHIMA T. *Systems of differential equations with regular singularities and their boundary value problems;* Annals of Math. 106 (1977), 145-200.

[20] KOMATSU H. *An introduction to the theory of hyperfunctions, Hyperfunctions and Pseudo-Differential Equations;* Lecture Notes in Math. 287 (1971), 3-40.

[21] MALGRANGE B. *Sur le points singuliers des équations différentielles;* L'Enseign. Math. 20 (1-2) (1974), 147-176.

[22] MELROSE R. *Normal self-intersections of the characteristic variety;* Bull. Amer. Math. Soc. 81 (1975), 939-940.

[23] MELROSE R. *Transformation methods for boundary value problems, Singularities in Boundary Value Problems;* D. Reidel P. Company (1981), 133-168.

[24] MIWA T. *Propagation of microanaliticity for solutions of pseudodifferential equations;* I, Publ. Res. Inst. Math. Sci. 10 (1975), 521-533.

[25] NOURRIGAT J.F. *Problèmes mixtes pour des systémes hyperboliques singuliers;* Astérisque 34-35 (1976), 251-261.

[26] ÔAKU T. *A canonical form of a system of microdifferential equations with non-involutory characteristic and branching of singularities;* Invent. Math. 65 (3) (1982), 491-525.

[27] ROBERTS G. *Uniqueness in the Cauchy problem for characteristic operators of Fuchsian type;* J. of Diff. Eq. 38 (1980), 374-392.

[28] TAHARA H. *Fuchsian type equations and fuchsian hyperbolic equations;* Japan J. Math. 5 (1979), 245-347.

[29] TAHARA H. *Singular hyperbolic systems*, III; On the Cauchy problem for fuchsian hyperbolic partial differential equations, J. Fac. Sci. Univ. Tokyo 27, I A, (1980), 465-507.

[30] TREVES F. *Second order Fuchsian elliptic equations and eigenvalue asymptotics;* Lecture Notes in Math. 459 (1975), 283-340.

SUBJECT INDEX

Bicharacteristic, 3, 135

Branching of singularities, 11, 134

Characteristic (manifold), 1

Canonical transformation, 105, 138

Conormal bundle, 2, 124

Fuchs (condition), 3, 25

Hamiltonian (vector field), 2, 115, 135

Indicial polynomial, 2, 115, 122, 132

Indicial equation, 2

Levi (condition) , 98, 134, 135

Microdistribution, 5, 93, 133

Non-involutive (manifold), 134

Operator. Pseudodifferential operator, 1

Fourier integral operator, 104, 139

Fuchsian operator, 6, 124

Fuchsian system, 1

Parametrix, 12, 14

Poisson bracket, 135

Propagation relation, 11, 13, 14

Symbol. Symbol of a pseudodifferential operatc

Principal symbol, 6, 97

Subprincipal symbol, 104

Symplectic (manifold), 8, 134

Wave front set (of a distribution), 1

Vol. 845: A. Tannenbaum, Invariance and System Theory: Algebraic and Geometric Aspects. X, 161 pages. 1981.

Vol. 846: Ordinary and Partial Differential Equations, Proceedings. Edited by W. N. Everitt and B. D. Sleeman. XIV, 384 pages. 1981.

Vol. 847: U. Koschorke, Vector Fields and Other Vector Bundle Morphisms – A Singularity Approach. IV, 304 pages. 1981.

Vol. 848: Algebra, Carbondale 1980. Proceedings. Ed. by R. K. Amayo. VI, 298 pages. 1981.

Vol. 849: P. Major, Multiple Wiener-Itô Integrals. VII, 127 pages. 1981.

Vol. 850: Séminaire de Probabilités XV. 1979/80. Avec table générale des exposés de 1966/67 à 1978/79. Edited by J. Azéma and M. Yor. IV, 704 pages. 1981.

Vol. 851: Stochastic Integrals. Proceedings, 1980. Edited by D. Williams. IX, 540 pages. 1981.

Vol. 852: L. Schwartz, Geometry and Probability in Banach Spaces. X, 101 pages. 1981.

Vol. 853: N. Boboc, G. Bucur, A. Cornea, Order and Convexity in Potential Theory: H-Cones. IV, 286 pages. 1981.

Vol. 854: Algebraic K-Theory. Evanston 1980. Proceedings. Edited by E. M. Friedlander and M. R. Stein. V, 517 pages. 1981.

Vol. 855: Semigroups. Proceedings 1978. Edited by H. Jürgensen, M. Petrich and H. J. Weinert. V, 221 pages. 1981.

Vol. 856: R. Lascar, Propagation des Singularités des Solutions d'Equations Pseudo-Différentielles à Caractéristiques de Multiplicités Variables. VIII, 237 pages. 1981.

Vol. 857: M. Miyanishi. Non-complete Algebraic Surfaces. XVIII, 244 pages. 1981.

Vol. 858: E. A. Coddington, H. S. V. de Snoo: Regular Boundary Value Problems Associated with Pairs of Ordinary Differential Expressions. V, 225 pages. 1981.

Vol. 859: Logic Year 1979–80. Proceedings. Edited by M. Lerman, J. Schmerl and R. Soare. VIII, 326 pages. 1981.

Vol. 860: Probability in Banach Spaces III. Proceedings, 1980. Edited by A. Beck. VI, 329 pages. 1981.

Vol. 861: Analytical Methods in Probability Theory. Proceedings 1980. Edited by D. Dugué, E. Lukacs, V. K. Rohatgi. X, 183 pages. 1981.

Vol. 862: Algebraic Geometry. Proceedings 1980. Edited by A. Libgober and P. Wagreich. V, 281 pages. 1981.

Vol. 863: Processus Aléatoires à Deux Indices. Proceedings, 1980. Edited by H. Korezlioglu, G. Mazziotto and J. Szpirglas. V, 274 pages. 1981.

Vol. 864: Complex Analysis and Spectral Theory. Proceedings, 1979/80. Edited by V. P. Havin and N. K. Nikol'skii, VI, 480 pages. 1981.

Vol. 865: R. W. Bruggeman, Fourier Coefficients of Automorphic Forms. III, 201 pages. 1981.

Vol. 866: J.-M. Bismut, Mécanique Aléatoire. XVI, 563 pages. 1981.

Vol. 867: Séminaire d'Algèbre Paul Dubreil et Marie-Paule Malliavin. Proceedings, 1980. Edited by M.-P. Malliavin. V, 476 pages. 1981.

Vol. 868: Surfaces Algébriques. Proceedings 1976–78. Edited by J. Giraud, L. Illusie et M. Raynaud. V, 314 pages. 1981.

Vol. 869: A. V. Zelevinsky, Representations of Finite Classical Groups. IV, 184 pages. 1981.

Vol. 870: Shape Theory and Geometric Topology. Proceedings, 1981. Edited by S. Mardešić and J. Segal. V, 265 pages. 1981.

Vol. 871: Continuous Lattices. Proceedings, 1979. Edited by B. Banaschewski and R.-E. Hoffmann. X, 413 pages. 1981.

Vol. 872: Set Theory and Model Theory. Proceedings, 1979. Edited by R. B. Jensen and A. Prestel. V, 174 pages. 1981.

Vol. 873: Constructive Mathematics, Proceedings, 1980. Edited by F. Richman. VII, 347 pages. 1981.

Vol. 874: Abelian Group Theory. Proceedings, 1981. Edited by R. Göbel and E. Walker. XXI, 447 pages. 1981.

Vol. 875: H. Zieschang, Finite Groups of Mapping Classes of Surfaces. VIII, 340 pages. 1981.

Vol. 876: J. P. Bickel, N. El Karoui and M. Yor. Ecole d'Eté de Probabilités de Saint-Flour IX – 1979. Edited by P. L. Hennequin. XI, 280 pages. 1981.

Vol. 877: J. Erven, B.-J. Falkowski, Low Order Cohomology and Applications. VI, 126 pages. 1981.

Vol. 878: Numerical Solution of Nonlinear Equations. Proceedings, 1980. Edited by E. L. Allgower, K. Glashoff, and H.-O. Peitgen. XIV, 440 pages. 1981.

Vol. 879: V. V. Sazonov, Normal Approximation – Some Recent Advances. VII, 105 pages. 1981.

Vol. 880: Non Commutative Harmonic Analysis and Lie Groups. Proceedings, 1980. Edited by J. Carmona and M. Vergne. IV, 553 pages. 1981.

Vol. 881: R. Lutz, M. Goze, Nonstandard Analysis. XIV, 261 pages. 1981.

Vol. 882: Integral Representations and Applications. Proceedings, 1980. Edited by K. Roggenkamp. XII, 479 pages. 1981.

Vol. 883: Cylindric Set Algebras. By L. Henkin, J. D. Monk, A. Tarski, H. Andréka, and I. Németi. VII, 323 pages. 1981.

Vol. 884: Combinatorial Mathematics VIII. Proceedings, 1980. Edited by K. L. McAvaney. XIII, 359 pages. 1981.

Vol. 885: Combinatorics and Graph Theory. Edited by S. B. Rao. Proceedings, 1980. VII, 500 pages. 1981.

Vol. 886: Fixed Point Theory. Proceedings, 1980. Edited by E. Fadell and G. Fournier. XII, 511 pages. 1981.

Vol. 887: F. van Oystaeyen, A. Verschoren, Non-commutative Algebraic Geometry, VI, 404 pages. 1981.

Vol. 888: Padé Approximation and its Applications. Proceedings, 1980. Edited by M. G. de Bruin and H. van Rossum. VI, 383 pages. 1981.

Vol. 889: J. Bourgain, New Classes of $\mathcal{L}^p$-Spaces. V, 143 pages. 1981.

Vol. 890: Model Theory and Arithmetic. Proceedings, 1979/80. Edited by C. Berline, K. McAloon, and J.-P. Ressayre. VI, 306 pages. 1981.

Vol. 891: Logic Symposia, Hakone, 1979, 1980. Proceedings, 1979, 1980. Edited by G. H. Müller, G. Takeuti, and T. Tugué. XI, 394 pages. 1981.

Vol. 892: H. Cajar, Billingsley Dimension in Probability Spaces. III, 106 pages. 1981.

Vol. 893: Geometries and Groups. Proceedings. Edited by M. Aigner and D. Jungnickel. X, 250 pages. 1981.

Vol. 894: Geometry Symposium. Utrecht 1980, Proceedings. Edited by E. Looijenga, D. Siersma, and F. Takens. V, 153 pages. 1981.

Vol. 895: J.A. Hillman, Alexander Ideals of Links. V, 178 pages. 1981.

Vol. 896: B. Angéniol, Familles de Cycles Algébriques – Schéma de Chow. VI, 140 pages. 1981.

Vol. 897: W. Buchholz, S. Feferman, W. Pohlers, W. Sieg, Iterated Inductive Definitions and Subsystems of Analysis: Recent Proof-Theoretical Studies. V, 383 pages. 1981.

Vol. 898: Dynamical Systems and Turbulence, Warwick, 1980. Proceedings. Edited by D. Rand and L.-S. Young. VI, 390 pages. 1981.

Vol. 899: Analytic Number Theory. Proceedings, 1980. Edited by M.I. Knopp. X, 478 pages. 1981.

MIX
Papier aus verantwortungsvollen Quellen
Paper from responsible sources
FSC® C105338

If you have any concerns about our products,
you can contact us on
ProductSafety@springernature.com

In case Publisher is established outside the EU,
the EU authorized representative is:
Springer Nature Customer Service Center GmbH
Europaplatz 3, 69115 Heidelberg, Germany

Printed by Libri Plureos GmbH
in Hamburg, Germany